Lecture Notes in Economics and Mathematical Systems

421

Chuangyin Dang

Triangulations
and Simplicial Methods

 Springer

Author

Chuangyin Dang
University of Auckland
Department of Engineering Science
Private Bag 92019
Auckland, New Zealand

ISBN-13: 978-3-540-58838-2   e-ISBN-13: 978-3-642-48775-0
DOI: 10.1007/ 978-3-642-48775-0

CIP data applied for

Typesetting: Camera ready by author
SPIN: 10486711   42/3140-543210 - Printed on acid-free paper

Acknowledgements

This monograph was written at the Center for Economic Research of Tilburg University in the Netherlands. It includes part of research in cooperation with my supervisor Dolf Talman. I deeply appreciate his guidance and collaboration. The monograph has benefited from the remarks of a number of people. I am very grateful to Gerard van der Laan for his helpful comments. I am also indebted to Eric van Damme and Walter Forster for their valuable suggestions. Moreover, I would like to express my appreciation to professors E.L. Allgower, R.W. Cottle, B.C. Eaves, K.G. Murty, R. Saigal, and M.J. Todd for stimulating discussions between us and to Kaizhou Chen, Xuchu He, and Zeke Wang for their encouragement. I want to thank all my colleagues, friends, and relatives for their help that I'll remember forever. Finally, I must say that I would never complete this monograph without understanding and support of my wife, Haijuan, and I would like to thank her for all of these.

Chuangyin Dang

Contents

1 Introduction   **1**

2 Preliminaries   **9**
    2.1   Notations 10
    2.2   Fixed Point Theorems 14
    2.3   Applications 15
       2.3.1   A Pure Exchange Economy 15
       2.3.2   A Finite n-Person Normal Form Game 17
       2.3.3   An Exchange Economy with Linear Production
           Technologies 18
       2.3.4   Convex Programming 18
       2.3.5   A Balanced Game without Side Payments 19

3 Existing Triangulations   **21**
    3.1   Existing Triangulations of S^n 21
    3.2   Existing Triangulations of R^n 27
    3.3   Existing Triangulations of Continuous Refinement of Grid
       Sizes of $(0, 1] \times S^n$ 32
    3.4   Existing Triangulations of Continuous Refinement of Grid
       Sizes of $(0, 1] \times R^n$ 40

4 The D_1-Triangulation of R^n   **45**
    4.1   The D_1-Triangulation of R^n 46
    4.2   Pivot Rules of the D_1-Triangulation 52
    4.3   The Number of Simplices of the D_1-Triangulation in a
       Unit Cube 54
    4.4   The Diameter of the D_1-Triangulation 56

4.5   The Average Directional Density of the D_1-Triangulation   58

5   The T_1-Triangulation of the Unit Simplex   65
5.1   The T_1-Triangulation   65
5.2   Pivot Rules of the T_1-Triangulation   73
5.3   Comparison of the Triangulations of the Unit Simplex . .   77

6   The D_1-Triangulation in Variable Dimension Algorithms on the Unit Simplex   79
6.1   The D_{v1}-Triangulation   81
6.2   Pivot Rules of the D_{v1}-Triangulation   88
6.3   The $(n+1)$-Ray Variable Dimension Method Based on the D_{v1}-Triangulation   92
6.4   The $(2^{n+1} - 2)$-Ray Variable Dimension Method Based on the D_{v1}-Triangulation   97

7   The D_1-Triangulation in Variable Dimension Algorithms on the Euclidean Space   105
7.1   The D_{v2}-Triangulation   106
7.2   Pivot Rules of the D_{v2}-Triangulation   113
7.3   The $2n$-Ray Variable Dimension Method Based on the D_1-Triangulation   118
7.4   The 2^n-Ray Variable Dimension Algorithm Based on the D_{v2}-Triangulation   121

8   The D_3-Triangulation for Simplicial Homotopy Algorithms   129
8.1   Definition of the D_3-Triangulation   130
8.2   Construction of the D_3-triangulation   132
8.3   Pivot Rules of the D_3-Triangulation   138
8.4   Comparison of Several Triangulations for Simplicial Homotopy Algorithms   142

9   The D_2-Triangulation for Simplicial Homotopy Algorithms   145
9.1   Construction of the D_2-Triangulation   146
9.2   Description of the D_2-Triangulation   148

9.3 Pivot Rules of the D_2-Triangulation 151

9.4 Description of the D_2^*-Triangulation 158

9.5 Pivot Rules of the D_2^*-Triangulation 162

9.6 Comparison of Several Triangulations for Simplicial Homotopy Algorithms 169

10 Conclusions **173**

Chapter 1

Introduction

The appearance of Brouwer's fixed point theorem in 1912 and its generalizations have resulted in a great breakthrough of a number of scientific research areas. Brouwer's theorem states that every continuous function from a compact and convex nonempty set into itself has a fixed point, i.e., an element which is mapped by the function into itself. However, the nonconstructive proofs of these fixed point theorems limited their further applications to real world problems. In 1967 Scarf gave the first elegant constructive proof of Brouwer's fixed point theorem on the unit simplex. The unit simplex is the subset of the Euclidean space in which all components of every point are nonnegative and sum up to one. From then on a significant development in computing fixed points was initiated.

Scarf's algorithm subdivides the unit simplex in a large number of the so-called primitive sets. Starting in a corner of the unit simplex, the algorithm generates a sequence of adjacent primitive sets until within a finite number of steps a primitive set is found that yields an approximate fixed point. However, operating with primitive sets needs an extremely high computer storage. In order to deal with this shortcoming, in 1968 Hansen developed an approach of choosing the primitive sets through a systematic way. But the pivot steps for moving from one primitive set to an adjacent one of Hansen's procedure were identical to the ones of a regular triangulation of the Euclidean space discovered by Freudenthal in 1942 and made operational on the unit simplex by Kuhn in 1960. This subdivision of the unit simplex into simplices was first used by Kuhn in algorithms for computing fixed points in 1968. Kuhn's original idea motivated many people to develop more efficient algorithms based on simplicial subdivisions. Since then almost all the algorithms for computing fixed points have been invented with triangulations. Actually, Kuhn proposed two simplicial algorithms on the unit simplex by himself. Kuhn's artificial start algorithm starts artificially outside the unit simplex and generates a sequence of adjacent full-dimensional simplices. Kuhn's variable dimension algorithm starts at one of vertices of the unit simplex and generates a sequence of adjacent simplices of varying dimension. Both algorithms of Kuhn terminate within a finite number of steps with a simplex yielding an approximate fixed point. The accuracy of approximation is completely determined by the grid sizes of the simplices. When the accuracy is not high enough one could restart the algorithm with a subdivision having simplices with smaller grid sizes. However, Kuhn's methods have the drawback that they discard all the information about the location of the fixed point obtained in the former implementation since they can only restart outside the unit simplex or at one of the vertices of the unit simplex. In order to deal with this drawback, in 1971 Merrill proposed an algorithm on the Euclidean space that starts at an arbitrary point on an artificial layer. This method is called the Sandwich algorithm, which was later rediscovered on the unit simplex by Kuhn and MacKinnon in 1975. To handle the same drawback, Eaves also proposed in 1972 an algorithm for computing fixed points on the unit simplex, called the simplicial homotopy algorithm, which was generalized to the Euclidean space by Eaves and Saigal in the same year. In both the Sandwich method and the simplicial homotopy algorithm, a homotopy function is constructed that deforms a trivial system into the original system. The fixed points of the homotopy function yield a curve which connects a fixed point of the trivial system with a fixed point of the original system. To follow this curve, the underlying space is subdivided into simplices and the homotopy function is linearized on each of these simplices. Then starting on the trivial layer, both the Sandwich method and the simplicial homotopy algorithm follow a piecewise linear path from the (unique) fixed point of the trivial system to an approximate fixed point of the original system. Within a finite number of steps the Sandwich method reaches a simplex on the original layer, which yields an approximate fixed point. If the accuracy is not good enough, one can restart the Sandwich method at the approximate fixed point obtained in the previous implementation with a smaller grid size of the simplices of the triangulation. The simplicial homotopy algorithm decreases continuously and automatically grid sizes of the simplices. It terminates as soon as the accuracy is high enough. So one doesn't need to make a restart of the simplicial homotopy algorithm in order to get a more accurate approximate fixed point. Both the Sandwich method and the simplicial homotopy method, however, need an extra dimension to start the procedure and in order to guarantee that the algorithms terminate within a finite number of steps. In addition, they both generate a sequence of full-dimensional simplices. Simplicial restart algorithms without an extra dimension were initiated by van der Laan and Talman in 1979. These so-called variable dimension algorithms start at an arbitrary point in the unit simplex or the Euclidean space and generate a sequence of adjacent simplices of varying dimension. They terminate within a finite number of steps with a simplex yielding an approximate fixed point. When the accuracy is not high enough, one can restart these algorithms at the approximate fixed point found in the previous implementation with a smaller grid size of the simplices of the triangulation. By now many simplicial variable dimension algorithms have been proposed both on the unit simplex and on the Euclidean space. These methods differ from each other in the number of rays along which the algorithms can leave the starting point. Simplicial algorithms might therefore be distinguished into three classes: 1) Sandwich algorithms, 2) simplicial homotopy algorithms, 3) simplicial variable dimension algorithms. Simplicial algorithms for computing fixed points have an advantage over Newton-type methods for their global convergence and capacity to deal with problems in which the underlying function is not differentiable or even not continuous. They have sometimes a shortcoming in converging slowly but they can be combined with Newton-type methods as has been suggested by Saigal and Todd in 1976.

It is obvious that triangulations play a basic role in simplicial algorithms for computing fixed points. The first triangulation of the Euclidean space was proposed by Freudenthal in 1942 for the topological proofs of some theorems. This triangulation was later made operational by Kuhn in 1960. The first application of Freudenthal's triangulation to simplicial algorithms was done by Kuhn for computing fixed points on the unit simplex in 1968. Utilizing Freudenthal's triangulation, Eaves constructed the first triangulation of continuous refinement of grid sizes of the unit simplex for simplicial homotopy algorithms in 1972. Eaves and Saigal proposed the first triangulation of continuous refinement of grid sizes of the Euclidean space in the same year. To improve simplicial algorithms, a new simplicial subdivision of the Euclidean space, the so-called Union Jack triangulation, was proposed by Todd in 1974, and in the same year Saigal also constructed a triangulation of the Euclidean space based on Freudenthal's triangulation. Numerical experience has shown that the efficiency of simplicial algorithms depends heavily on the underlying triangulation. In order to compare various simplicial subdivisions, Saigal proposed in 1974 as the first theoretical measure of efficiency of triangulations the number of simplices of a triangulation in a unit cube. In 1975 Saigal, Solow and Wolsey introduced as a measure of efficiency of triangulations the diameter of a triangulation. According to these two measures they discovered that Freudenthal's triangulation, the Union Jack triangulation and Saigal's triangulation have the same number of simplices in the unit cube and that in the n-dimensional Euclidean space the diameter of Freudenthal's triangulation and the diameter of the Union Jack triangulation are the same, whereas the diameter of Saigal's triangulation is equal to or greater than n times that of Freudenthal's triangulation. They therefore concluded that Freudenthal's triangulation and the Union Jack triangulation are superior to Saigal's triangulation. This theoretical result is confirmed by some numerical tests. In 1976 Todd investigated systematically the efficiency of triangulations. He proposed as a theoretical measure of efficiency of triangulations the average directional density of a triangulation. Depending on this measure Todd showed that Freudenthal's triangulation and the Union Jack triangulation are better than Saigal's triangulation. To improve simplicial homotopy algorithms, Todd constructed further in 1976 a new triangulation of continuous refinement of grid sizes based on the Union Jack triangulation. Motivated by measures of efficiency of triangulations, van der Laan and Talman found in 1980 an optimal transformation such that Freudenthal's triangulation under this transformation has the smallest average directional density.

Their transformed Freudenthal's triangulation provides a more efficient simplicial subdivision for their original variable dimension algorithm on the Euclidean space. Since the triangulations proposed by Eaves, by Eaves and Saigal and by Todd have a fixed refinement factor of two, it stimulated to construct triangulations of continuous and arbitrary refinement of grid sizes. In 1980 van der Laan and Talman proposed a triangulation of continuous and arbitrary refinement of grid sizes for simplicial homotopy algorithms based on Freudenthal's triangulation. At the same time Shamir discovered independently a similar triangulation. In 1982 Kojima and Yamamoto found a series of triangulations of continuous and arbitrary refinement of grid sizes based on Freudenthal's triangulation and on the Union Jack triangulation. After these developments, in 1984 Eaves gave a comprehensive investigation on triangulations for simplicial homotopy algorithms. These triangulations have made a big improvement of simplicial homotopy algorithms. In order to get rid of the artificial level, in 1987 Broadie and Eaves combined simplicial variable dimension algorithms together with simplicial homotopy algorithms. It seems to be natural that not every triangulation is as suitable as another one for a simplicial algorithm and more important that for different simplicial algorithms the same triangulation is not always most efficient. From this view point, in 1987 Doup and Talman constructed a triangulation based on Freudenthal's triangulation for use in the simplicial variable dimension algorithm of van der Laan and Talman on the unit simplex. By utilizing the location of the starting point their triangulation has caused many advantages in efficiency and led to several new simplicial variable dimension algorithms on the unit simplex and the Cartesian product of the unit simplices called the simplotope. All together, in order to improve simplicial algorithms and to construct more efficient triangulations for simplicial algorithms, a large number of triangulations have been constructed for over the last two decades.

The history of development of simplicial algorithms and their applications is only about twenty years, but a huge amount of literature has appeared. Allgower and Georg presented an excellent survey on these topics in 1980. Above we only gave a historical description of simplicial algorithms on the unit simplex and the Euclidean space. Simplicial algorithms on simplotopes also have been developed recently. For this we refer to Doup's comprehensive description of simplicial algorithms on simplotopes in 1988.

Numerical tests have shown that the underlying triangulation influences considerably the efficiency of simplicial algorithms. Hence, it is very significant to propose better triangulations for simplicial algorithms. Our research interest in triangulations for simplicial algorithms was stimulated by this fact. In this monograph a new triangulation of the Euclidean space is constructed. It is called the D_1-triangulation. This triangulation induces a simplicial subdivision of every unit cube. It is shown that the D_1-triangulation is superior to all other well-known triangulations for simplicial algorithms according to measures of efficiency of triangulations such as the number of simplices in a unit cube, the diameter, and the average directional density. Secondly, a new triangulation of the unit simplex is presented. It is called the T_1-triangulation and is a combination of the D_1-triangulation and the Union Jack triangulation. It is suitable only for one of variable dimension methods on the unit simplex. Next, we consider how to incorporate in general the D_1-triangulaton in variable dimension algorithms on the unit simplex. A version of the D_1-triangulation is developed such that it induces according to the D_1-triangulation a simplicial subdivision of each of the subsets, into which a simplicial variable dimension algorithm subdivides the unit simplex. In addition, we discuss how to use the D_1-triangulation in simplicial variable dimension algorithms on the Euclidean space. Therefore, another version of the D_1-triangulation is given such that it induces according to the D_1-triangulation a simplicial subdivision of each of the subsets, into which a simplicial variable dimension algorithm subdivides the Euclidean space. It is also considered how to use the D_1-triangulation in simplicial homotopy algorithms. Therefore, a new triangulation of continuous refinement of grid sizes is constructed. It is called the D_3-triangulation. This triangulation is superior to the other triangulations for simplicial homotopy algorithms. However, the D_3-triangulation has a fixed refinement factor of two. In order to loose this limition, a triangulation of continuous and arbitrary refinement of grid sizes is given. It is called the D_2-triangulation. But this triangulation doesn't induce the D_3-triangulation as its special case. Therefore, another triangulation of continuous refinement of grid sizes is presented, called the D_2^*-triangulation, such that its refinement factors can be chosen as arbitrary even integers while it induces the D_3-triangulation as its special case.

This monograph is organized as follows. Chapter 2 gives the basic concepts and several applications of fixed point theorems. Existing triangulations of the unit simplex and the Euclidean space are described in Chapter 3. The D_1-triangulation is introduced in Chapter 4. The T_1-triangulation is presented in Chapter 5. How to use the D_1-triangulation in variable dimension algorithms on the unit simplex and on the Euclidean space is considered in Chapter 6 and 7, respectively. The D_3-triangulation is proposed in Chapter 8. The D_2-triangulation and the D_2^*-triangulation are described in Chapter 9. Finally, Chapter 10 concludes with some computational results.

Chapter 2

Preliminaries

As an elegant mathematical tool for proving the existence of a solution for some significant mathematical problems, Brouwer's fixed point theorem and its generalizations have applications to economics, game theory, networks, transportation, engineering, and many other fields. Brouwer's fixed point theorem simply says that a continuous function, mapping a convex and compact nonempty set into itself, has a fixed point. One of its generalizations, Kakutani's fixed point theorem, contains a more extensive situation. In scientific research areas a great number of problems can be reduced or are equivalent to the existence problem of a fixed point. Fixed point theorems were initially proved by nonconstructive approaches. Thus they were only able to show the existence of fixed points. It was Scarf who proposed the first method to be able to yield an approximate fixed point. Since then the computation of fixed points and applications have been developed considerably. Most algorithms for computing fixed points triangulate or subdivide the set, on which the problem is defined, into simplices and are therefore called simplicial algorithms. This monograph is focused on triangulations and their applications in simplicial algorithms on the unit simplex and on the Euclidean space. This chapter introduces in the first section some basic notations, states in the second section some relevent fixed point and related existence theorems, and gives in the last section some applications to economics, game theory, and mathematical programming.

2.1   Notations

In this section some basic notations used in this monograph are given. The k-dimensional Euclidean space is denoted by R^k. We write the real line as R. The set of all nonnegative vectors in R^k is represented by R^k_+. The n-dimensional unit simplex is equal to the set S^n defined by

$$S^n = \left\{ x \in R^{n+1}_+ \,\middle|\, \sum_{j=0}^n x_j = 1 \right\}.$$

For a given finite set A, $\#(A)$ or $|A|$ denotes the number of elements in A. We say that a set in the Euclidean space is compact if it is both closed and bounded and that a set is convex if any convex combination of two points in the set belongs to it. The convex hull of a set is equal to the intersection of all convex sets containing this set. The convex hull of a set A is denoted by conv(A). The affine hull of a set is equal to the intersection of all affine subspaces containing this set. The affine hull of a set A is denoted by aff(A). For a given set A, its dimension means the dimension of the linear subspace parallel to the affine hull of A and is denoted by dim(A).

Definition 2.1.1. The vectors y^0, y^1, $\cdots$, y^k in R^n are affinely independent if

$$\sum_{j=0}^k \alpha_j y^j = 0 \text{ and } \sum_{j=0}^k \alpha_j = 0$$

imply $\alpha_j = 0$ for all j.

The convex hull of $k + 1$ affinely independent vectors y^0, y^1, $\cdots$, y^k is called a k-dimensional simplex or a k-simplex. The vectors y^0, y^1, $\cdots$, y^k are called vertices of the simplex. Let σ denote a k-simplex. A face of σ is a simplex that is the convex hull of some of vertices of σ. A face τ of the k-simplex σ is a facet if dim(τ) $= k - 1$. A facet τ of the k-simplex σ is called the facet opposite to vertex y of σ if y is the vertex of σ not being a vertex of τ.

Definition 2.1.2. Let C be a convex subset of R^n with dim(C) $= m$. G is a triangulation or a simplicial subdivision of the set C if

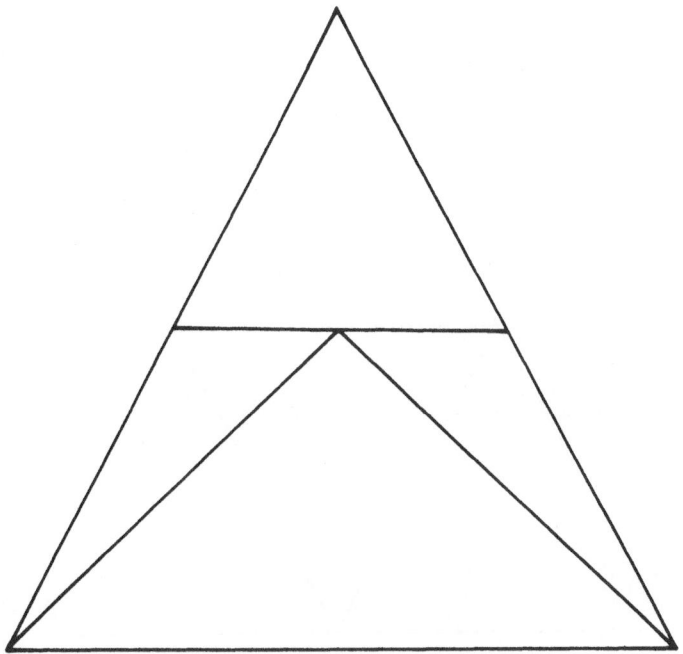

Figure 2.1: Not a triangulation

1. G is a collection of m-dimensional simplices,

2. C is equal to the union of all simplices in G,

3. for any σ^1 and σ^2 in G, the intersection of σ^1 and σ^2 is either empty or a common face of both σ^1 and σ^2,

4. every x in C has a neighborhood meeting only a finite number of simplices in G.

Example 2.1.3. The following figures show the geometrical interpretation of the definition of a triangulation of a set. The collection of simplices in Figure 2.1 is not a triangulation of S^2, whereas the collection of simplices in Figure 2.2 is a triangulation of S^2.

A very important property of a triangulation G of a convex set C is that every facet of a simplex in G either lies in the boundary of C and is a facet of no other simplex in G or it doesn't lie in the boundary of C and is a facet of exactly one other simplex in G.

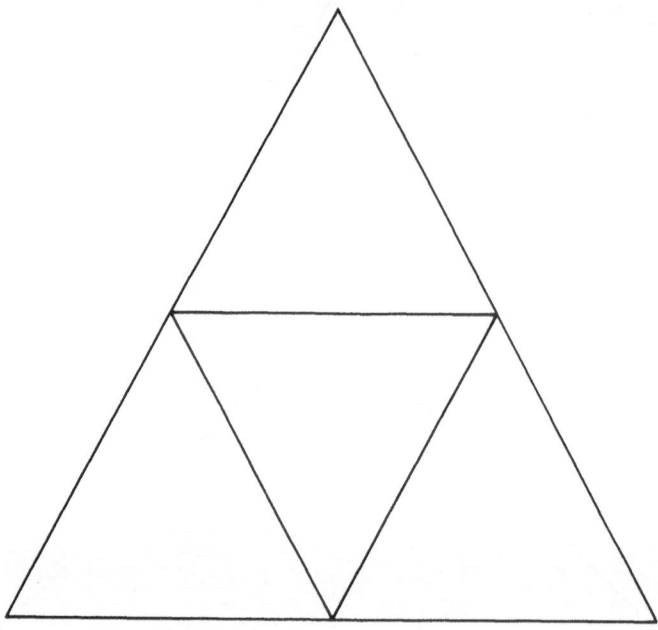

Figure 2.2: A triangulation

For a given vector $x \in R^n$, we define its 1-norm, 2-norm, ∞-norm by

$$\|x\|_1 = \sum_{i=1}^{n} |x_i|,$$

$$\|x\|_2 = \sqrt{x_1^2 + x_2^2 + \cdots + x_n^2},$$

$$\|x\|_\infty = \max_{1 \leq i \leq n} |x_i|,$$

respectively. Let G be a triangulation of a convex set C. For a simplex $\sigma \in G$, the diameter of σ is equal to

$$\text{diam}(\sigma) = \max \{\|x - y\|_\infty|\ x,\ y \in \sigma\}.$$

Sometimes we also call the diameter of a simplex the grid size. The mesh of G is equal to

$$\text{mesh}(G) = \sup \{\text{diam}(\sigma) \mid \sigma \in G\}.$$

The n-dimensional unit cube is denoted by U^n, i.e.,

$$U^n = \{x \in R^n \mid 0 \leq x_i \leq 1 \text{ for } i = 1, 2, \cdots, n\}.$$

Let G be a triangulation of U^n. The number of simplices of G in U^n is denoted by $\mathcal{N}(G)$.

The set ∂U^n represents the boundary of U^n. Let τ and τ' be two different facets of G in ∂U^n. Furthermore, let $\sigma_0, \sigma_1, \cdots, \sigma_m$ be a sequence of adjacent simplices, i.e., $\sigma_i \cap \sigma_{i+1}$ is a facet of both σ_i and σ_{i+1} for $i = 0, 1, \cdots, m-1$. If τ is a facet of σ_0 and τ' a facet of σ_m, then we call $\sigma_0, \sigma_1, \cdots, \sigma_m$ a path of length $m+1$ from τ to τ'. The distance between τ and τ' is defined to be the length of the shortest path between them.

Definition 2.1.4. The diameter of G is equal to the maximal distance between any two facets in ∂U^n.

The diameter of G is denoted by $\mathcal{D}(G)$.

The set $[x, y]$ denotes the line segment between x and y in R^n. Let G be a triangulation of R^n.

Definition 2.1.5. The average directional density of G is equal to

$$\text{avg}_{\|d\|_2=1} \lim_{t \to \infty} \lim_{r \to \infty} \text{avg}_{\|x\|_2 \le r} \frac{1}{t} \#(\{\sigma \in G \mid \sigma \cap [x, x+td] \ne \emptyset\}),$$

where t and r are in R, the inner average is taken with respect to the Lebesgue measure on the ball of radius r, and the outer average is taken over a random vector d uniformly distributed on the surface of the ball of radius one, and where the limits are assumed to exist.

The average directional density of G is denoted by $\mathcal{A}(G)$. It measures simply the expected number of simplices of G met by an average unit length of randomly placed lines.

Let C be a nonempty convex subset of R^n. A function $f : C \to R$ is called convex on C if

$$f(\lambda x^1 + (1 - \lambda)x^2) \le \lambda f(x^1) + (1 - \lambda)f(x^2) \tag{2.1}$$

for any different x^1 and x^2 in C and any $0 \le \lambda \le 1$. A function $f : C \to R$ is called quasi-convex on C if

$$f(\lambda x^1 + (1 - \lambda)x^2) \le \max\left\{f(x^1), f(x^2)\right\} \tag{2.2}$$

for any different x^1 and x^2 in C and any $0 \le \lambda \le 1$. A function $f : C \to R$ is called strictly convex on C if the strict inequality in inequality 2.1 holds for any $0 < \lambda < 1$. A function $f : C \to R$ is called strictly quasi-convex on C if the strict inequality in inequality 2.2 holds for any $0 < \lambda < 1$. A function $f : C \to R$ is called concave, quasi-concave, strictly concave, and strictly quasi-concave on C if $-f$ is convex, quasi-convex, strictly convex, and strictly quasi-convex on C, respectively.

2.2   Fixed Point Theorems

Let C be a compact and convex nonempty subset of R^n. Let $f : C \to C$ be a continuous function. In 1912 Brouwer obtained the following conclusion.

Theorem 2.2.1. There exists a point x^* in C such that $f(x^*) = x^*$.

We call a point which is mapped by f to itself a fixed point of f in C.

Let X be a subset of R^n and let $\mathcal{Y}$ denote the set of subsets of R^m. Furthermore, let $F : X \to \mathcal{Y}$ be a point to set mapping such that $F(x)$ is not empty for every $x \in X$.

Definiton 2.2.2. F is upper semicontinuous at a point $x^0 \in X$ if for every open set H containing $F(x^0)$ there exists a neighborhood $N(x^0)$ of x^0 such that $F(x)$ is contained in H for every $x \in N(x^0)$. F is upper semicontinuous on X if $F(x)$ is compact and upper semicontinuous at every $x \in X$.

Let F be an upper semicontinuous mapping from C to the set of nonempty convex subsets of C. In 1941 Kakutani proved the following result.

Theorem 2.2.3. There exists a point $x^* \in C$ such that $x^* \in F(x^*)$.

Let $f : C \to R^n$ be a continuous function.

Definition 2.2.4. A point $x^* \in C$ is a stationary point of f if

$$(x - x^*)^\mathsf{T} f(x^*) \leq 0$$

for all $x \in C$.

From **Theorem 2.2.3** we can prove the following results.

Corollary 2.2.5. There exists a stationary point x^* of f in C.

A continuous function $f : S^n \to R^{n+1}$ is called a complementary function on S^n if $x^\mathsf{T} f(x) = 0$ for all $x \in S^n$.

Corollary 2.2.6. If f is a complementary function on S^n, then there exists at least one point x^* such that $f(x^*) \leq 0$.

A continuous function $f : S^n \to R^{n+1}$ is called an excess demand function if

1. $x^\mathsf{T} f(x) = 0$ for all $x \in S^n$,

2. $x_i = 0$ implies $f_i(x) \geq 0$ for all i.

Corollary 2.2.7. If f is an excess demand function on S^n, then there exists at least one point $x^* \in S^n$ such that $f(x^*) = 0$.

2.3 Applications

In this section some applications of the exsistence theorems stated in the previous section are discussed to such as economics, game theory, and mathematical programming.

2.3.1 A Pure Exchange Economy

Let there be $n + 1$ commodities, indexed by $i \in N_0$ defined by the set $\{0, 1, \cdots, n\}$, traded among m consumers, indexed by $j \in M$ defined by the set $\{1, 2, \cdots, m\}$. Consumer j has an initial endowment of the commodities equal to $w^j \in R_+^{n+1}$, where w_i^j is his endowment of commodity i. We assume that $w_i^j > 0$ for all i and j. Under a price vector $p \in R_+^{n+1} \backslash \{0\}$, the budget set of consumer j is equal to the set

$$B^j(p) = \left\{ x \in R_+^{n+1} \mid p^\top x \le p^\top w^j \text{ and } x \le \sum_{j=1}^{m} w^j \right\}.$$

Let a utility function $u^j : R_+^{n+1} \to R$ represent the preference ordering of consumer j over all commodity bundles. It is assumed that u^j is continuous, strictly monotone and strictly quasi-concave. For a given price $p \in R_+^{n+1} \backslash \{0\}$, the demand $d^j(p)$ of consumer j is determined by his preference and his budget constraint as follows: $d^j(p)$ of consumer j

$$\text{maximize } u^j(x)$$
$$\text{subject to } x \in B^j(p).$$

The solution $d^j(p)$ is uniquely determined for a given price $p \in R_+^{n+1} \backslash \{0\}$ and d^j is continuous on $R_+^{n+1} \backslash \{0\}$. From the strict monotonicity of u^j it is clear that consumer j will use up all his income. Thus $p^\top d^j(p) = p^\top w^j$ for $p \in R_+^{n+1} \backslash \{0\}$. Further, d^j is homogeneous of degree zero, i.e., $d^j(\lambda p) = d^j(p)$ for $\lambda > 0$. Let $d(p) = \sum_{j=1}^{m} d^j(p)$ denote the aggregate demand of the consumers under price $p \in R_+^{n+1} \backslash \{0\}$ and let $w = \sum_{j=1}^{m} w^j$ denote the aggregate endowment of the consumers. Then d is continuous on $R_+^{n+1} \backslash \{0\}$, $p^\top d(p) = p^\top w$ for $p \in R_+^{n+1} \backslash \{0\}$ (Walras' law), and d is homogeneous of degree zero. Thus one can normalize a price $p \in R_+^{n+1} \backslash \{0\}$ so that the sum of its components is equal to one. For a given price $p \in S^n$, when $p_i = 0$, from the strict monotonicity of u^j it is obvious that $d_i(p) \ge w_i$. Therefore, the function $z : S^n \to R^{n+1}$, defined by $z(p) = d(p) - w$, is an excess demand function on the unit simplex S^n.

Definition 2.3.1. A price vector p^* in S^n is an equilibrium price vector if $z(p^*) = 0$.

From **Corollary 2.2.7** the following result is concluded.

Theorem 2.3.2. There exists an equilibrium price vector in S^n.

2.3.2 A Finite n-Person Normal Form Game

Let $\Gamma = (\Phi_1, \Phi_2, \cdots, \Phi_n, R_1, R_2, \cdots, R_n)$ be a finite n-person normal form game, where Φ_i is the set of pure strategies of player i and $R_i : \Pi_{j=1}^n \Phi_j \to R$ is the payoff function for $i = 1, 2, \cdots, n$. A mixed strategy s_i of player i is a distribution over all his pure strategies. The number $s_i^{\phi_i}$ denotes the probability with which player i takes his pure strategy $\phi_i \in \Phi_i$ when he plays a mixed strategy s_i. Let

$$S_i = \left\{ s_i \,\middle|\, \sum_{\phi_i \in \Phi_i} s_i^{\phi_i} = 1, s_i^{\phi_i} \geq 0 \text{ for all } \phi_i \in \Phi_i \right\}$$

for $i = 1, 2, \cdots, n$. Define $\Phi = \Pi_{j=1}^n \Phi_j$ and $S = \Pi_{j=1}^n S_j$. When a mixed strategy s is played, the payoff of player i is equal to

$$R_i(s) = \sum_{\phi \in \Phi} \Pi_{j=1}^n s_j^{\phi_j} R_i(\phi)$$

for $i = 1, 2, \cdots, n$.

Definition 2.3.3. A Nash equilibrium is a mixed strategy s^* such that

$$\sum_{\phi \in \Phi} (s_i^{\phi_i} - s_i^{\phi_i *}) \Pi_{j \neq i} s_j^{\phi_j *} R_i(\phi) \leq 0$$

for all i.

Let $\phi_{-i} = (\phi_1, \cdots, \phi_{i-1}, \phi_{i+1}, \cdots, \phi_n)$ and $\Phi_{-i} = \Pi_{j \neq i} \Phi_j$. For $i = 1, 2, \cdots, n$, let $f_{\phi_i} : S \to R$ be defined by

$$f_{\phi_i}(s) = \sum_{\phi_{-i} \in \Phi_{-i}} \Pi_{j \neq i} s_j^{\phi_j} R_i(\phi)$$

for all $\phi_i \in \Phi_i$. Define the mapping

$$f = (f_{\phi_1} : \phi_1 \in \Phi_1; f_{\phi_2} : \phi_2 \in \Phi_2; \cdots; f_{\phi_n} : \phi_n \in \Phi_n)^{\mathsf{T}}.$$

Then s^* is a Nash equilibrium if and only if s^* is a stationary point of f in S. Thus from **Corollary 2.2.5** the following conclusion is derived.

Theorem 2.3.4. Every finite n-person normal form game has a Nash equilibrium.

2.3.3   An Exchange Economy with Linear Production Technologies

Let there be $n+1$ commodities indexed by $i \in \{0, 1, \cdots, n\}$, a finite number of consumers, and m firms indexed by $j \in \{1, 2, \cdots, m\}$. Suppose that the consumers have an initial aggregate endowment $w \in R_+^{n+1}$. Assume that the aggregate demand of the consumers, $d(p)$, is convex-valued and upper semi-continuous on the price space S^n and satisfies Walras' law, i.e., for $x \in d(p)$, $p^\top x = p^\top w$ for $p \in S^n$. The production activity of firm j at a unit level is a vector $a^j \in R^{n+1}$ whose negative components correspond to the inputs and whose positive components correspond to the outputs. Let $y \in R_+^m$ denote an activity level vector with component y_j the activity level of firm j for $j = 1, 2, \cdots, m$. Let A be equal to the matrix $(a^1, a^2, \cdots, a^m)$. Then the vector Ay is the aggregate net input-output vector for an activity level vector $y \in R_+^m$. For a price $p \in S^n$, $A^\top p$ is equal to the unit level profit of the activities. Assume that it is not possible to produce without input, i.e., $Ay \geq 0$ and $y \geq 0$ imply $y = 0$.

Definition 2.3.5. A vector $(p^*, y^*)^\top \in S^n \times R_+^m$ is an equilibrium if

1. $Ay^* + w \in d(p^*)$,

2. $A^\top p^* \leq 0$.

From **Theorem 2.2.3** the following result is induced.

Theorem 2.3.6. There exists an equilibrium vector in $S^n \times R_+^m$.

2.3.4   Convex Programming

Let us consider the problem

$$(CP) \quad \begin{array}{c} \text{minimize } f(x) \\ \text{subject to } g_i(x) \leq 0, \ i = 1, 2, \cdots, m, \end{array}$$

where f and g_i, $i = 1, 2, \cdots, m$, are functions from R^n to R. Assume that these functions are convex on R^n.

Definition 2.3.7. A subgradient of a convex function h at a point $x \in R^n$ is equal to a vector ξ such that

$$h(y) - h(x) \geq \xi^T(x - y)$$

for all $y \in R^n$.

Let $\partial h(x)$ denote the set of all the subgradients of h at $x \in R^n$. Then $\partial h(x)$ is called the subdifferential of h at $x \in R^n$. It is obvious that ∂h is upper semicontinuous on R^n. Let $q(x) = \max_{1 \leq i \leq m} g_i(x)$ for $x \in R^n$. Let $I(x)$ be equal to $\{i \mid q(x) = g_i(x)\}$ for $x \in R^n$. Then $\partial q(x)$ is equal to the convex hull of the set $\cup_{i \in I(x)} \partial g_i(x)$. Since f and g_i, $i = 1, 2, \cdots, m$, are convex, ∂f and ∂q are convex-valued and upper semicontinuous on R^n. Let the mapping p defined by

$$p(x) = \begin{cases} \{x\} - \partial f(x), & \text{if } q(x) < 0, \\ \{x\} - \text{conv}(\partial f(x) \cup \partial q(x)), & \text{if } q(x) = 0, \\ \{x\} - \partial q(x), & \text{if } q(x) > 0 \end{cases}$$

for $x \in R^n$. Then p is convex-valued and upper semicontinuous on R^n.

Theorem 2.3.8. If there exists a point $x \in R^n$ with $q(x) < 0$, then the set of the fixed points of p coincides with the set of optimal solutions of problem (CP).

From **Theorem 2.2.3** the next result is induced, see [133].

Corollary 2.3.9. If there exists a point $x \in R^n$ with $q(x) < 0$ and if the set $\{x \mid q(x) \leq 0\}$ is bounded, then problem (CP) has a solution.

2.3.5 A Balanced Game without Side Payments

Let there be n players indexed by i, $i \in N$ defined by the set $\{1, 2, \cdots, n\}$. Let 2^N denote the set of all nonempty subsets of N. An element in 2^N is called a coalition. A cooperative game is a pair (N, f), where f is a mapping from 2^N to the set of subsets of R^n. The set $f(S)$ denotes the payoffs that the players in coalition $S \in 2^N$ are able to get by their cooperation regardless of the actions of the players outside of the coalition. Let R^S denote the $|S|$-dimensional subspace of R^n with co-ordinates indexed by the elements in S. If $x \in R^n$ and $S \in 2^N$, then x^S will denote the projection of x on R^S. The following assumptions are imposed on f. For $S \in 2^N$, the set $f(S)$ satisfies that

1. for $y \in R^n$, if $x \in f(S)$ and $x_i = y_i$ for all $i \in S$, then $y \in f(S)$,

2. for $y \in R^n$, if $x \in f(S)$ and $y \le x$, then $y \in f(S)$,

3. $f(S)$ is closed,

4. $\left\{ x^S \mid x \in f(S) \right\}$ is nonempty and bounded from above.

Without loss of generality we assume that the set $f(\{i\})$ has been normalized to the half space $\{x \mid x_i \le 0\}$ for $i = 1, 2, \cdots, n$, and that the other $f(S)$'s have been shifted accordingly.

Definition 2.3.10. The core of the game (N, f) is equal to the set

$$C(N, f) = \left\{ x \in f(N) \; \middle| \; \begin{array}{l} \text{there are no } S \in 2^N \text{ and } y \in f(S) \\ \text{such that } y_i > x_i \text{ for all } i \in S \end{array} \right\}.$$

Let B be a collection of nonempty subsets of 2^N. Define B_i to be equal to the set $\{S \in B \mid i \in S\}$. The set B is called balanced if there exist nonnegative numbers μ_S, $S \in B$, such that $\sum_{S \in B_i} \mu_S = 1$ for all $i \in S$.

Definition 2.3.11. A game (N, f) is balanced if for every balanced set B, the intersection of $f(S)$ over all $S \in B$ is contained in $f(N)$.

The following conclusion is induced from **Theorem 2.2.3**, see [167].

Theorem 2.3.12. Every balanced game has a nonempty core.

Chapter 3

Existing Triangulations

Triangulations or simplicial subdivisions appeared originally in the category of topology. Now they play a basic role in simplicial algorithms for computing fixed points. Numerical experience has shown that simplicial algorithms depend heavily on the underlying triangulation. In order to develop more efficient simplicial algorithms, a great number of triangulations have been proposed. The introduction of new triangulations is caused by the fact that for different simplicial algorithm not always the same triangulation is the most efficient one and each of simplicial algorithms needs a triangulation suitable to itself. This chapter introduces in Section 1 some of the most well-known triangulations of the unit simplex S^n, gives in Section 2 some triangulations of R^n, and describes in Section 3 and 4 some triangulations of continuous refinement of grid sizes, respectively.

3.1 Existing Triangulations of S^n

In the sequel, let N denote the index set $\{1, 2, \cdots, n\}$ and N_0 the index set $\{0, 1, \cdots, n\}$. For $i = 0, 1, \cdots, n$, the vector u^i denotes the i-th unit vector in R^{n+1}, i.e., $u^i = (0, \cdots, 1, \cdots, 0)^\top$ with the one on the i-th place. For $i = 0, 1, \cdots, n-1$, the $(n+1)$-vector q^{i+1} is defined by $q^{i+1} = u^i - u^{i+1}$. We call a vector $s = (s_1, s_2, \cdots, s_n)^\top$ a sign vector if $s_i \in \{-1, +1\}$ for all i.

The most well-known triangulation of the n-dimensional unit simplex S^n is the Q-triangulation proposed by Freudenthal in [51]. It has applications to simplicial algorithms such as the Sandwich method and the $(n+1)$-ray variable dimension algorithm on the unit simplex. The simplices of the Q-triangulation can be described as follows. Let a positive integer m be given. We call m^{-1} the grid size. Next, a vector $y \in S^n$ is chosen such that every component of y is a non-negative multiple of $1/m$ with $y_n \neq 0$. Further, take a permutation $\pi = (\pi(1), \pi(2), \cdots, \pi(n))$ of the elements of N such that if $y_k = 0$ with $k > 0$ then $i > j$, where $\pi(i) = k$ and $\pi(j) = k+1$.

Definition 3.1.1. For the vector y and the permutation π given as above, the vectors $y^0, y^1, \cdots, y^n$ are given as follows:

$$y^0 = y,$$
$$y^i = y^{i-1} + q^{\pi(i)}/m, \ i = 1, 2, \cdots, n.$$

Clearly, the vectors $y^0, y^1, \cdots, y^n$ are affinely independent points in S^n. Their convex hull is therefore an n-simplex and is denoted by $Q(y, \pi)$. Let Q represent the collection of simplices $Q(y, \pi)$ for all y and π given as above. Then Q is a simplicial subdivision of S^n, called the Q-triangulation with grid size m^{-1}. It is illustrated in Figure 3.1 for $n = 2$ and $m = 4$.

Secondly, we describe the J_1'-triangulation of S^n given by Todd in [176]. The simplices of the J_1'-triangulation are defined as follows. Let a positive integer m be given. Then a vector $y \in S^n$ is chosen such that my_0 is odd and my_i is even for $1 \leq i < n$. A sign vector $s \in R^n$ is taken such that for all $k > 0$, if $y_k = 0$ and $s_k = 1$ then $s_{k+1} = 1$. Take a permutation $\pi = (\pi(1), \pi(2), \cdots, \pi(n))$ of the elements of N such that when $y_k = 0$, if $s_k = 1$ then $i > j$ and if $s_{k+1} = -1$ then $i < j$, where $\pi(i) = k$ and $\pi(j) = k+1$.

Definition 3.1.2. For y, π and s given as above, the vectors $y^0, y^1, \cdots, y^n$ are given as follows:

$$y^0 = y,$$
$$y^i = y^{i-1} + s_{\pi(i)}q^{\pi(i)}/m, \ i = 1, 2, \cdots, n.$$

Let J_1' represent the collection of simplices $J_1'(y, \pi, s)$ that are the convex hull of $y^0, y^1, \cdots, y^n$, as obtained from **Definition 3.1.2**, for all y,

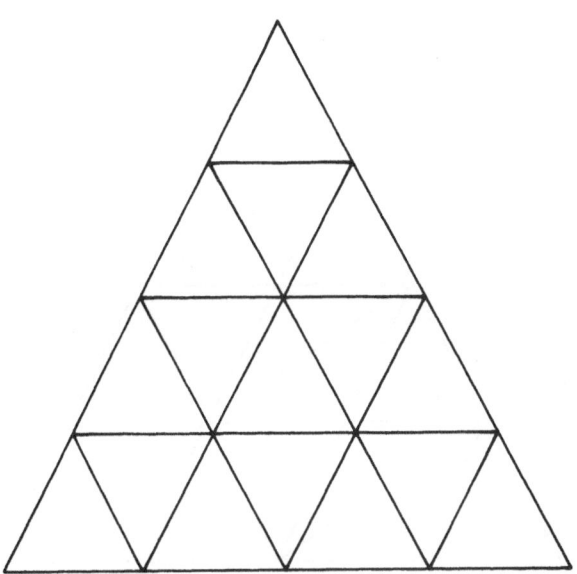

Figure 3.1: The Q-triangulation for $n = 2$ and $m = 4$.

π and s given as above. Then J'_1 is a simplicial subdivision of S^n, called the J'_1-triangulation with grid size m^{-1}. It is illustrated in Figure 3.2 for $n = 2$ and $m = 4$.

Next, we give the U-triangulation of the affine hull of S^n presented by van der Laan and Talman in [110]. It has a smaller average directional density than all other known triangulations of S^n but it is only able to subdivide the affine hull of S^n into simplices. The simplices of the U-triangulation are defined as follows. Let u denote the $(n + 1)$-vector such that all its components are equal to one. For $i = 1, 2, \cdots, n$, we define the $(n+1)$-vector t^i by $t^i = u - (n+1)u^i$, i.e., $t^i = (1, \cdots, -n, \cdots, 1)^\top$ with the $-n$ on the i-th place. Let a positive integer m be given. Then choose an $(n + 1)$-vector y in the affine hull of S^n such that $y = u/(n+1) + m^{-1} \sum_{i=1}^n \alpha_i t^i$, where α_i, $i = 1, 2, \cdots, n$, are integers. Take a permutation $\pi = (\pi(1), \pi(2), \cdots, \pi(n))$ of the elements of N.

Definition 3.1.3. For y and π given as above, the vectors y^0, y^1, $\cdots$,

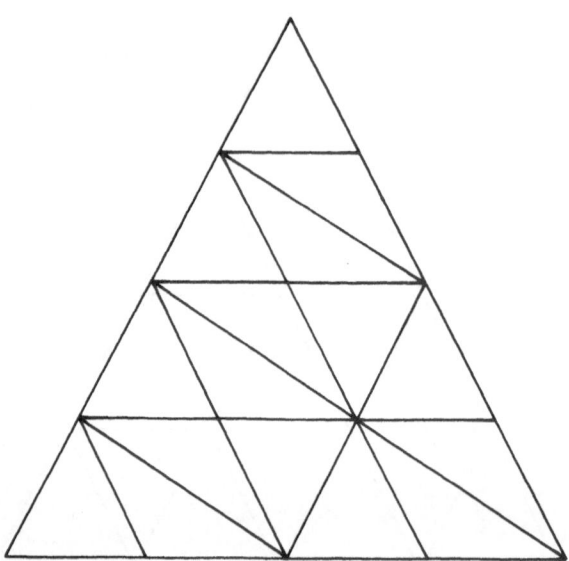

Figure 3.2: The J_1'-triangulation for $n = 2$ and $m = 4$.

y^n are given as follows:

$$y^0 = y,$$
$$y^i = y^{i-1} + t^{\pi(i)}/m, \quad i = 1, 2, \cdots, n.$$

Let U represent the collection of simplices $U(y, \pi)$ that are the convex hull of $y^0, y^1, \cdots, y^n$, as obtained from **Definition 3.1.3**, for all y and π given as above. Then U is a simplicial subdivision of the affine hull of S^n, called the U-triangulation with grid size m^{-1}. It is illustrated in Figure 3.3 for $n = 2$.

Finally, we introduce the V-triangulation of S^n proposed by Doup and Talman in [27]. This triangulation has stimulated to develop several new simplicial variable dimension algorithms for computing economic equilibria. The simplices of the V-triangulation are defined as follows. Let a vector $x^0 \in S^n$ and a positive integer m be given. The projection vector $v(H)$ of x^0 on the subset

$$S^n(H) = \{x \in S^n \mid x_i = 0 \text{ for all } i \notin H\}$$

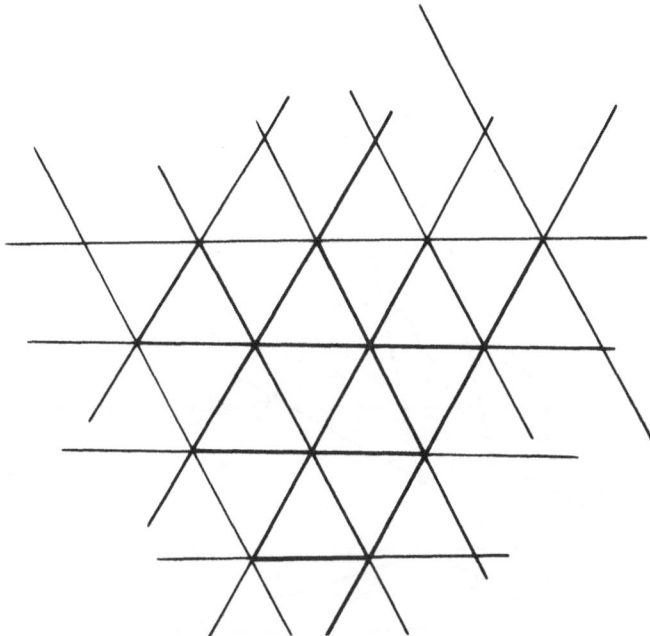

Figure 3.3: The U-triangulation for $n = 2$.

of S^n, $H \subset N_0$ and $H \neq \emptyset$, is defined by

$$
v_i(H) = \begin{cases} 0 & \text{if } i \notin H, \\ (1 - \sum_{j \in H} x_j^0)/(\sum_{j \in H} x_j^0 + |H_0|) & \text{if } i \in H_0, \\ x_i^0(1 + |H_0|)/(\sum_{j \in H} x_j^0 + |H_0|) & \text{otherwise} \end{cases}
$$

for $i = 0, 1, \cdots, n$, where H_0 is the set of the indices $j \in H$ such that $x_j^0 = 0$. When H is empty, we define $v(H) = x^0$. Take a permutation $p = (p(1), p(2), \cdots, p(n))$ of n elements out of N_0. Let T denote the set of indices in the permutation p. Further, let us define the $(n+1)$-vector $w^{p(i)}$ by

$$
w^{p(i)} = v(\{p(1), \cdots, p(i)\}) - v(\{p(1), \cdots, p(i-1)\})
$$

for $i = 1, 2 \cdots, n$. Next, choose a vector

$$
y = x^0 + \sum_{i=1}^{n} \lambda_{p(i)} w^{p(i)}/m,
$$

where $\lambda_{p(i)}$, $i = 1, 2, \cdots, n$ are integers such that $m - 1 \geq \lambda_{p(1)} \geq \cdots \geq \lambda_{p(n)} \geq 0$. Take a permutation $\pi = (\pi(1), \pi(2), \cdots, \pi(n))$ of elements in

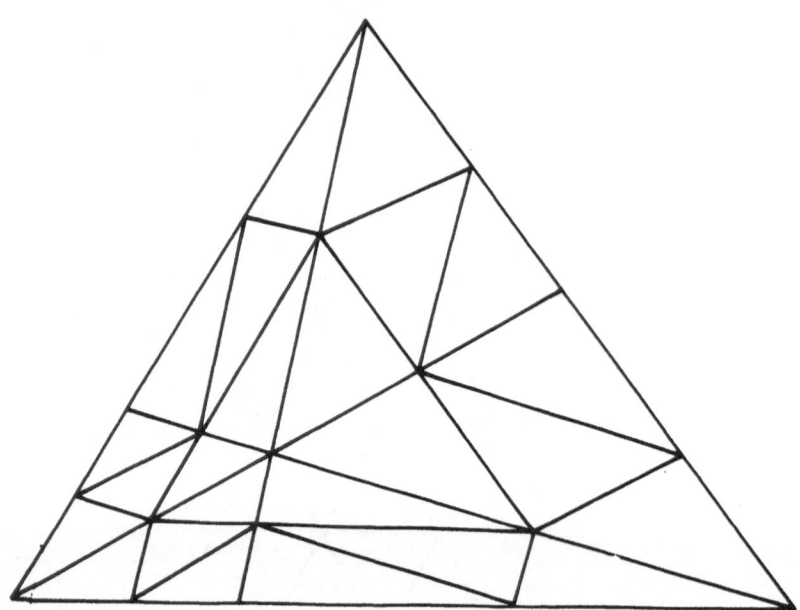

Figure 3.4: The V-triangulation for $n = 2$ and $m = 2$.

T such that when $\pi(i) = p(k)$ and $\pi(j) = p(k-1)$ for some $1 < k \le n$, $i > j$ if $\lambda_{p(k-1)} = \lambda_{p(k)}$.

Definition 3.1.4. For y and π given as above, the vectors y^0, y^1, $\cdots$, y^n are given as follows:

$$y^0 = y,$$
$$y^i = y^{i-1} + w^{\pi(i)}/m, \ i = 1, 2, \cdots, n.$$

Let V represent the collection of simplices $V(y, \pi)$ that are the convex hull of y^0, y^1, $\cdots$, y^n, as obtained from **Definition 3.1.4**, for all y, π and p given as above. Then V is a simplicial subdivision of S^n, called the V-triangulation with grid size m^{-1}. It is illustrated in Figure 3.4 for $n = 2$ and $m = 2$.

We remark that also other triangulations of S^n have been proposed such as the iterated barycentric subdivision used by Shapley in [167] and its closely related triangulation proposed by Zangwill in [198], and so on. However, these triangulations are not appropriate to use in simplicial algorithms and are therefore not considered in this monograph.

3.2 Existing Triangulations of R^n

Let N again denote the index set $\{1, 2, \cdots, n\}$. The vector u^i represents the i-th unit vector in R^n for $i = 1, 2, \cdots, n$.

The K_1-triangulation is the best known simplicial subdivision of R^n and was proposed by Freudenthal in [51]. It has applications in all simplicial algorithms. We describe the simplices of the K_1-triangulation as follows. A vector $y \in R^n$ is chosen such that every component of y is an integer. Take a permutation $\pi = (\pi(1), \pi(2), \cdots, \pi(n))$ of the elements in N.

Definition 3.2.1. For y and π given as above, the vectors y^0, y^1, $\cdots$, y^n are given as follows:

$$y^0 = y,$$
$$y^i = y^{i-1} + u^{\pi(i)}, \; i = 1, 2, \cdots, n.$$

Let K_1 denote the collection of simplices $K_1(y, \pi)$ that are the convex hull of y^0, y^1, $\cdots$, y^n, as obtained from **Definition 3.2.1**, for all y and π given as above. Then K_1 is a simplicial subdivision of R^n, called the K_1-triangulation. It is illustrated in Figure 3.5 for $n = 2$ and $n = 3$.

Next we introduce the J_1-triangulation given by Todd in [179]. A nice property of this simplicial subdivision is that it subdivides R^n symmetrically into simplices. The simplices of the J_1-triangulation are defined as follows. First, a vector y is chosen in R^n such that every component of y is odd. Then a permutation $\pi = (\pi(1), \pi(2), \cdots, \pi(n))$ of the elements in N and a sign vector $s = (s_1, s_2, \cdots, s_n)^\mathsf{T}$ are taken.

Definition 3.2.2. For y, π and s given as above, the vectors y^0, y^1, $\cdots$, y^n are given as follows:

$$y^0 = y$$
$$y^i = y^{i-1} + s_{\pi(i)} u^{\pi(i)}, \; i = 1, 2, \cdots, n.$$

Let J_1 denote the collection of simplices $J_1(y, \pi, s)$ that are the convex hull of y^0, y^1, $\cdots$, y^n, as obtained from **Definition 3.2.2**, for all y, π and s given as above. Then J_1 forms a simplicial subdivision of R^n, called the J_1-triangulation. It is illustrated in Figure 3.6 for $n = 2$ and

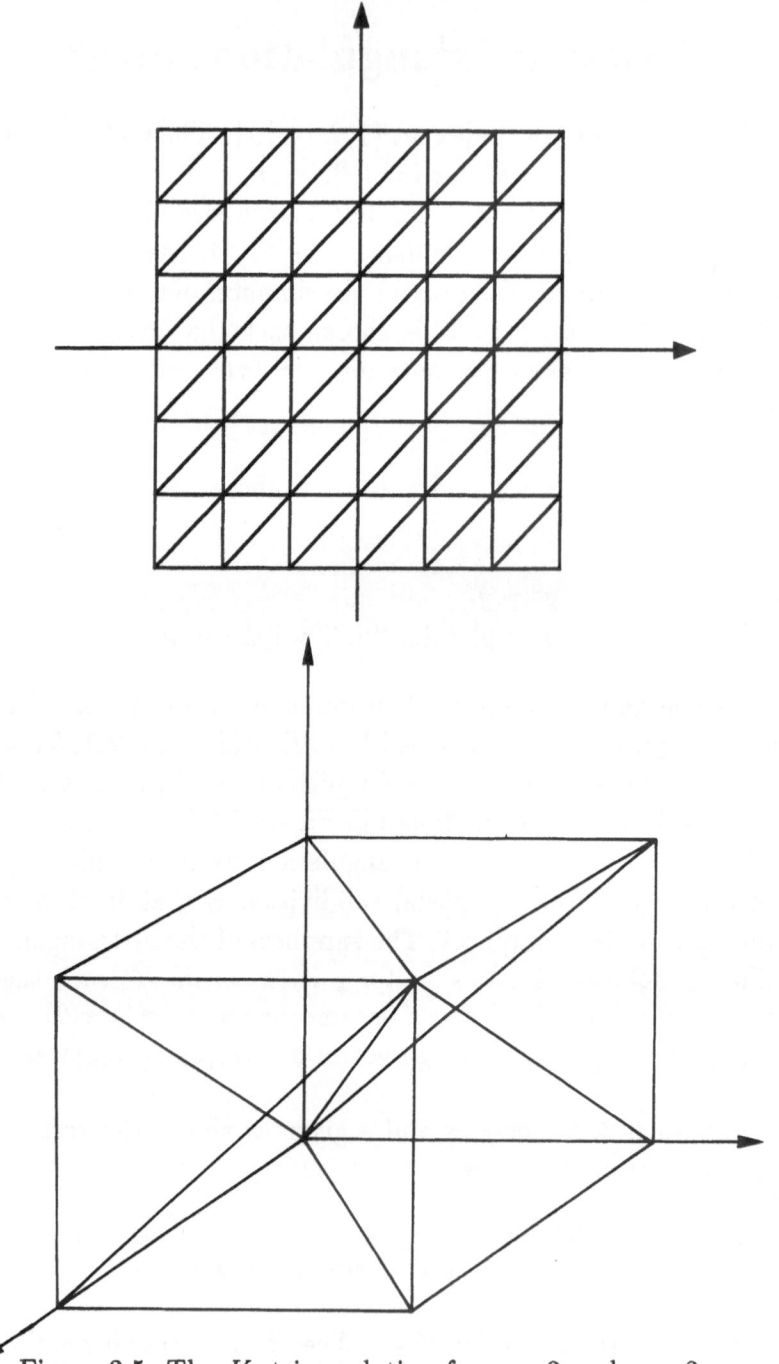

Figure 3.5: The K_1-triangulation for $n = 2$ and $n = 3$.

$n = 3$. Note that when all components of y are even, J_1 also subdivides R^n into simplices.

A less efficient simplicial subdivision of R^n is the so called H_1-triangulation described by Saigal in [150]. It can be obtained with a linear transformation from the K_1-triangulation and is related to the Q-triangulation of S^n. We give the simplices of the H_1-triangulation as follows. Let us define the n-vector p^i by $p^i = (0, \cdots, 0, -1, 1, 0, \cdots, 0)^T$ with -1 on the i-th place for $i = 1, 2, \cdots, n-1$, and $p^n = (0, \cdots, 0, -1)^T$. Then a vector y is chosen such that every component of y is an integer. Take a permutation $\pi = (\pi(1), \pi(2), \cdots, \pi(n))$ of the elements in N.

Definition 3.2.3. For y and π given as above, the vectors $y^0, y^1, \cdots, y^n$ are given as follows:

$$y^0 = y,$$
$$y^i = y^{i-1} + p^{\pi(i)}, \ i = 1, 2, \cdots, n.$$

Let H_1 denote the collection of simplices $H_1(y, \pi)$ that are the convex hull of $y^0, y^1, \cdots, y^n$, as obtained from **Definition 3.2.3**, for all y and π given as above. Then H_1 is a simplicial subdivision of R^n, called the H_1-triangulation. It is illustrated in Figure 3.7 for $n = 2$ and $n = 3$.

The next simplicial subdivision we introduce has been used in the 2^n-ray and $(3^n - 1)$-ray variable dimension algorithms on R^n. It is called the K'-triangulation proposed by Todd in [180]. The simplices of the K'-triangulation are defined as follows. Choose a vector y such that every component of y is an integer. Then define $I^+(y) = \{i \in N \mid y_i > 0\}$, $I^-(y) = \{i \in N \mid y_i < 0\}$, and $I^0(y) = \{i \in N \mid y_i = 0\}$. A sign vector is chosen such that $s_i = 1$ for $i \in I^+(y)$, $s_i = -1$ for $i \in I^-(y)$ and $s_i \in \{-1, +1\}$ for $i \in I^0(y)$. Take a permutation $\pi = (\pi(1), \pi(2), \cdots, \pi(n))$ of the elements in N.

Definition 3.2.4. For y, π and s given as above, the vectors $y^0, y^1, \cdots, y^n$ are given as follows:

$$y^0 = y$$
$$y^i = y^{i-1} + s_{\pi(i)} u^{\pi(i)}, \ i = 1, 2, \cdots, n.$$

Let K' denote the collection of simplices $K'(y, \pi, s)$ that are the convex hull of $y^0, y^1, \cdots, y^n$, as obtained from **Definition 3.2.4**, for all y, π

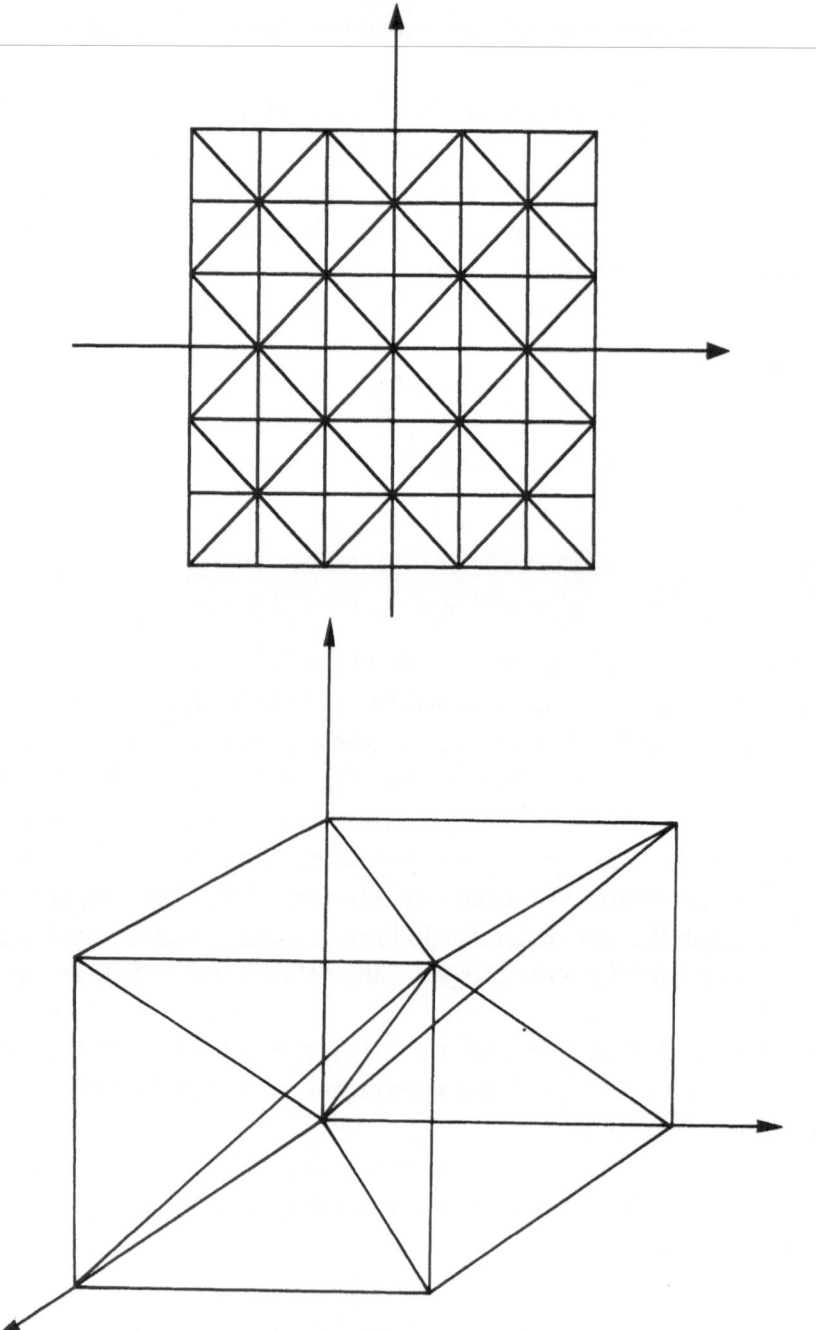

Figure 3.6: The J_1-triangulation for $n = 2$ and $n = 3$.

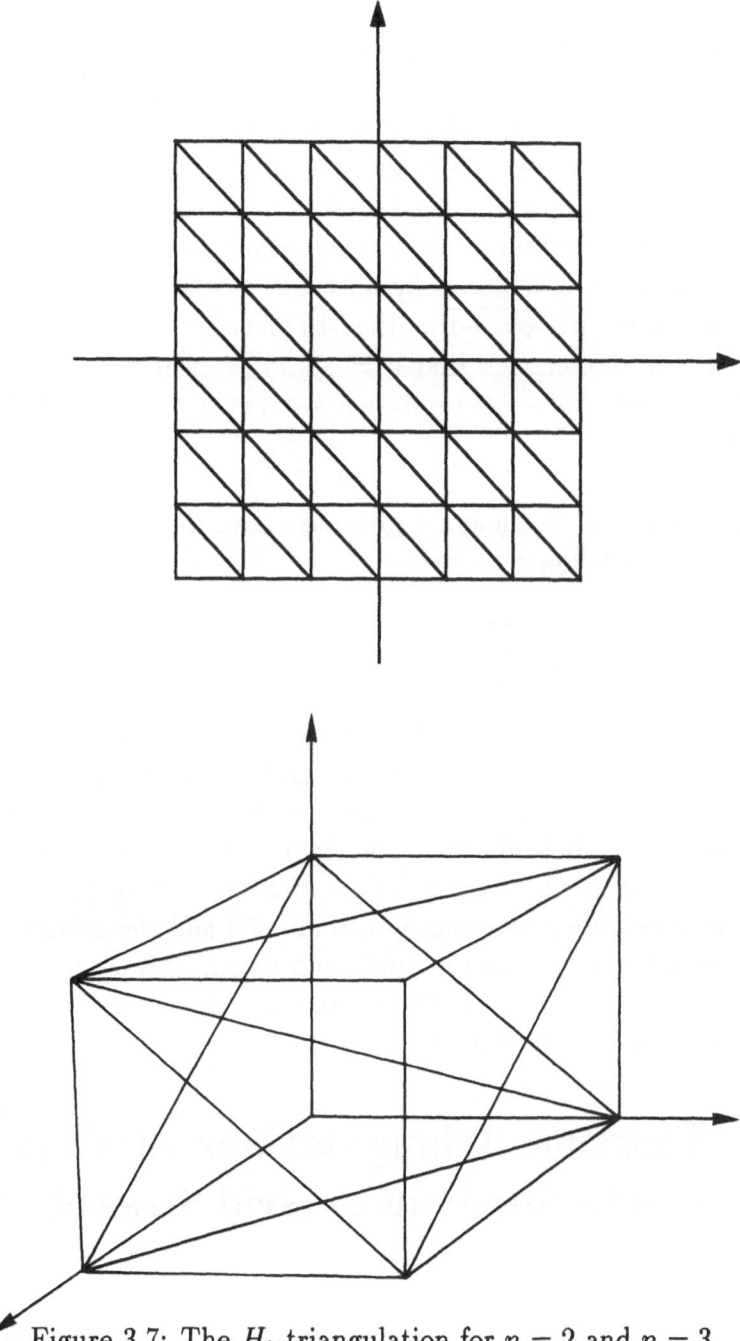

Figure 3.7: The H_1-triangulation for $n = 2$ and $n = 3$.

and s given as above. Then K' forms a simplicial subdivision of R^n, called the K'-triangulation. It is illustrated in Figure 3.8 for $n = 2$ and $n = 3$.

Finally, we introduce the A^*-triangulation presented by van der Laan and Talman in [110]. It has a smaller average directional density than other known simplicial subdivisions of R^n. It can be used in the $(n + 1)$-ray variable dimension algorithm on the Euclidean space. Let us define the n-vector a^i by $a^i = (-1, \cdots, n + \sqrt{n+1}, \cdots, -1)^\mathsf{T}$ with the number $n + \sqrt{n+1}$ on the i-th place for $i = 1, 2, \cdots, n$. Then a vector y is chosen such that $y = \sum_{i=1}^{n} \alpha_i a^i$, where α_i is an integer for all i. Finally, take a permutation $\pi = (\pi(1), \pi(2), \cdots, \pi(n))$ of the elements in N.

Definition 3.2.5. For y and π given as above, the vectors y^0, y^1, $\cdots$, y^n are given as follows:

$$y^0 = y,$$
$$y^i = y^{i-1} + a^{\pi(i)}, \; i = 1, 2, \cdots, n.$$

Let A^* denote the collection of simplices $A^*(y, \pi)$ that are the convex hull of y^0, y^1, $\cdots$, y^n, as obtained from **Definition 3.2.5**, for all y and π given as above. Then A^* is a simplicial subdivision of R^n, called the A^*-triangulation. It is illustrated in Figure 3.9 for $n = 2$ and $n = 3$.

Note that also other triangulations of R^n have been proposed such as the triangulations obtained by Lee in [123] and the middle cut triangulations given by Sallee in [160], and so on. However, since these triangulations are complicated to use in simplicial algorithms, they are not introduced in this monograph.

3.3 Existing Triangulations of Continuous Refinement of Grid Sizes of $(0, 1] \times S^n$

In a triangulation of continuous refinement of grid sizes of the set $(0, 1] \times S^n$, for some sequence of positive integers k_0, k_1, $\cdots$, the set S^n on level 1 is subdivided into simplices according to some triangulation with

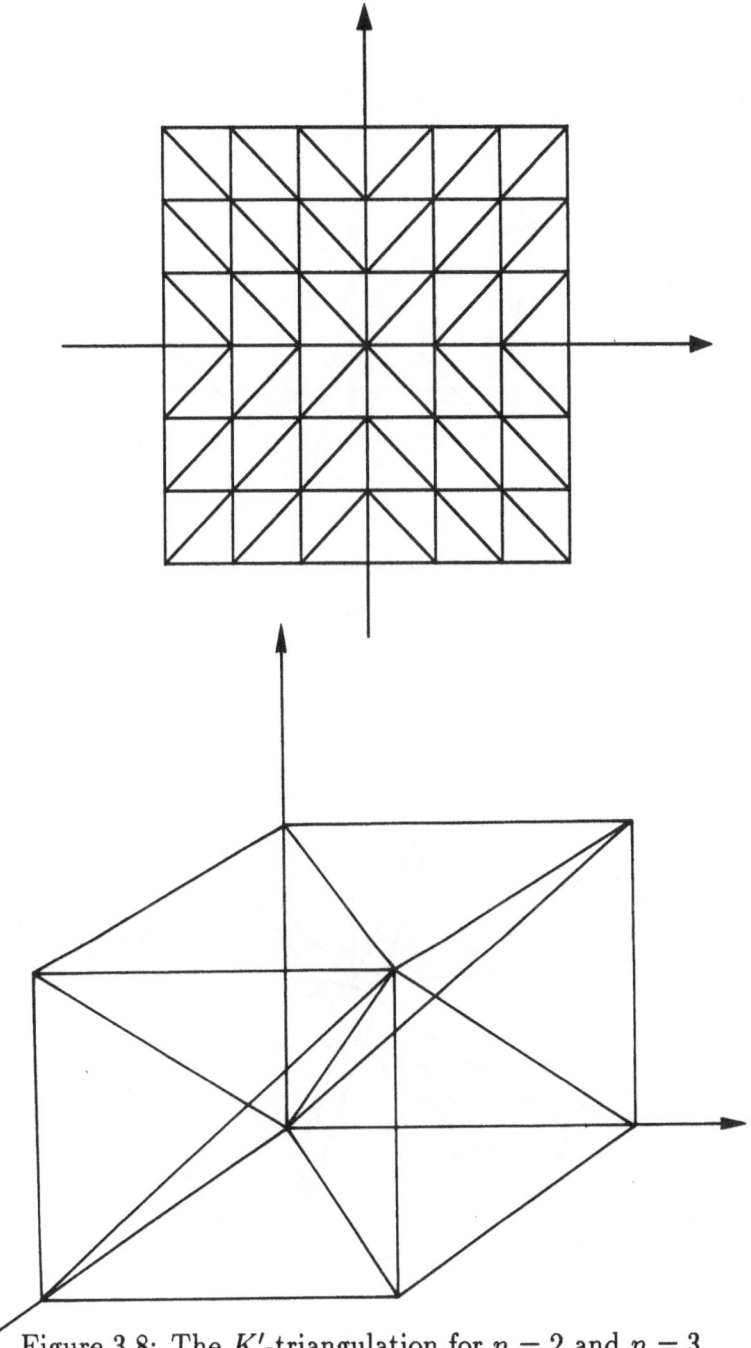

Figure 3.8: The K'-triangulation for $n = 2$ and $n = 3$.

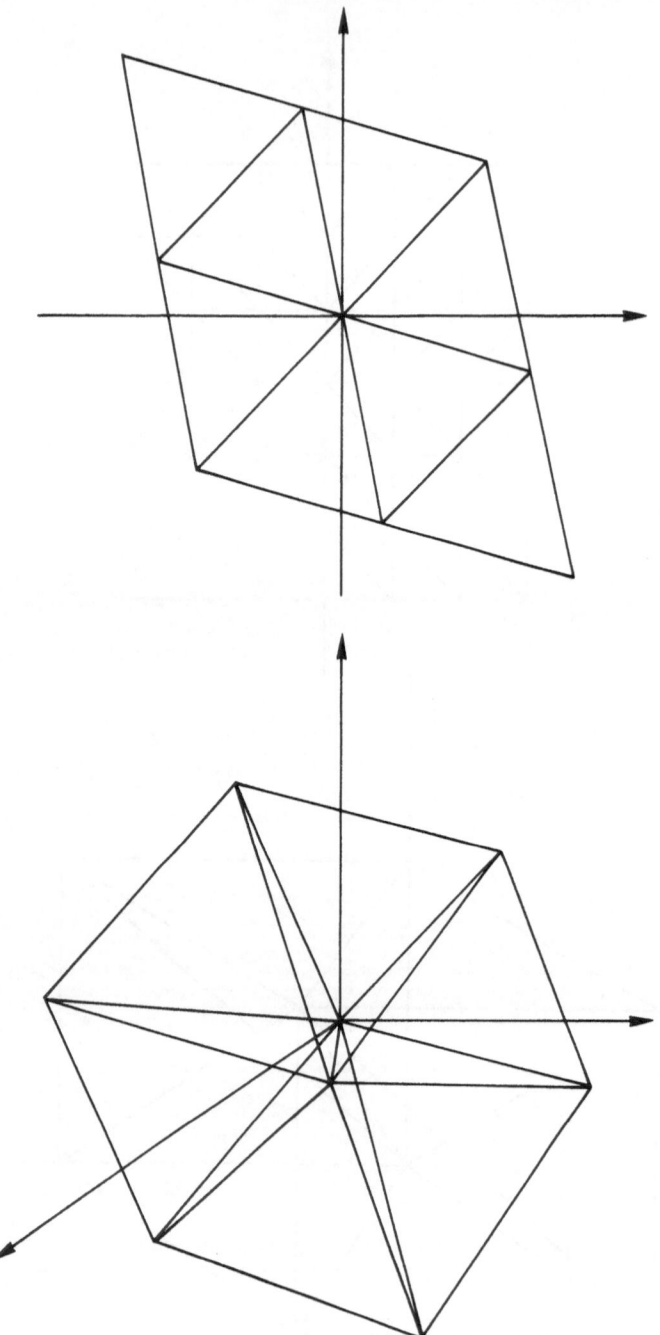

Figure 3.9: The A^*-triangulation for $n = 2$ and $n = 3$.

grid size m^{-1} for some positive integer m, whereas on level d_0/d_i where $d_0 = m$ and $d_i = k_{i-1}d_{i-1}$ for $i = 1, 2, \cdots$, the set S^n is subdivided into simplices according to the same triangulation with a grid size being a factor k_{i-1} smaller than the one on level d_0/d_{i-1}. The first of these triangulations have a fixed refinement factor of grid sizes equal to 2.

Let N_0 again denote the index set $\{0, 1, \cdots, n\}$. Let the $(n+2)$-vector u^i denote the i-th unit vector in R^{n+2} for $i = 0, 1, \cdots, n+1$. Then we define the $(n+2)$-vector h^i by $h^i = u^{i+1} - u^i$ for $i = 1, 2, \cdots, n$.

As follows we introduce the K_3'-triangulation of continuous refinement of grid sizes of $(0, 1] \times S^n$. It is the first triangulation for simplicial homotopy algorithms on S^n and was proposed by Eaves in [34]. Let a positive integer m be given. Then take a permutation $\pi = (\pi(0), \pi(1), \cdots, \pi(n))$ of the elements in N_0. Let g denote the integer such that $\pi(g) = 0$. For an integer $k \geq 1$, choose a vector y such that $y_0 = 2^{-k}$ and

$$y = y_0 u^0 + u^1 + y_0 \sum_{i=1}^{n} \lambda_i h^i / m,$$

where λ_i are integers for all i such that $2^k m \geq \lambda_1 \geq \cdots \geq \lambda_n \geq 0$, $\lambda_{\pi(i)} < \lambda_{\pi(j)}$ if both $\pi(i) > \pi(j)$ and $i < j$, $\lambda_{\pi(i)} \leq 2^k m - 1$ for all $0 \leq i \leq g - 1$, and where $\lambda_{\pi(i)}$ is odd for all $g + 1 \leq i \leq n$. Then define

$$w_{\pi(i)} = \begin{cases} 0 & \text{if } \lambda_{\pi(i)} \text{ is odd,} \\ 1/m & \text{otherwise} \end{cases}$$

for $i = 0, 1, \cdots, g - 1$.

Definition 3.3.1. For y and π given as above, the vectors $y^{-1}, y^0, \cdots, y^n$ are given as follows:

$$y^{-1} = y,$$
$$y^i = y^{i-1} + y_0 h^{\pi(i)}/m, \quad i = 0, \cdots, g - 1,$$
$$y^g = y^{g-1} - y_0(\sum_{j=0}^{g-1} w_{\pi(j)} h^{\pi(j)} + \sum_{j=g+1}^{n} h^{\pi(j)}/m) + y_0 u^0,$$
$$y^i = y^{i-1} + 2y_0 h^{\pi(i)}/m, \quad i = g + 1, \cdots, n.$$

Let K_3' denote the collection of simplices $K_3'(y, \pi)$ that are the convex hull of $y^{-1}, y^0, \cdots, y^n$, as obtained from **Definition 3.3.1**, for all y

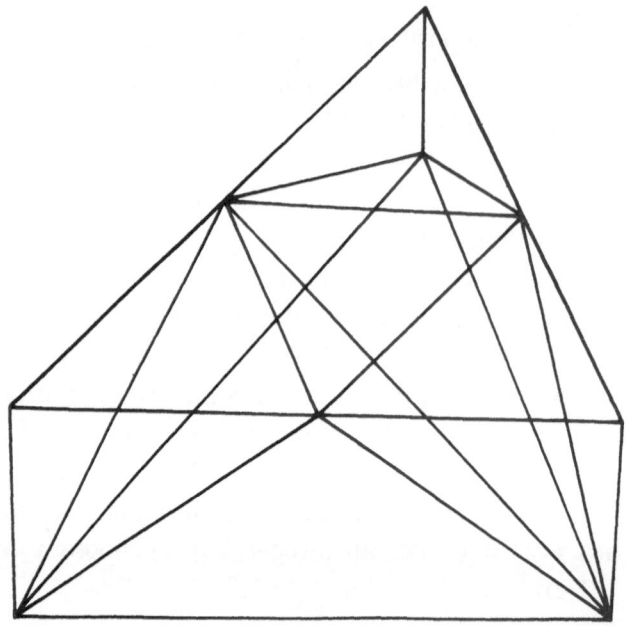

Figure 3.10: The K_3'-triangulation for $n = 2$ and $m = 1$.

and π given as above. Then K_3' is a simplicial subdivision of continuous refinement of grid sizes of $(0,1] \times S^n$, called the K_3'-triangulation. It is illustrated in Figure 3.10 for $n = 2$ and $m = 1$.

Next we introduce the J_3'-triangulation of continuous refinement of grid sizes of $(0,1] \times S^n$. This triangulation was proposed by Todd in [176]. The simplices of the J_3'-triangulation are defined as follows. Let a positive integer m be given. For an integer $k \geq 1$, choose a vector y such that $y_0 = 2^{-k}$ and

$$y = y_0 u^0 + u^1 + y_0 \sum_{i=1}^{n} \lambda_i h^i / m,$$

where $2^k m \geq \lambda_1 \geq \cdots \geq \lambda_n \geq 0$ and λ_i is odd for all $1 \leq i \leq n$. Let us define

$$t_i = \begin{cases} -1 & \text{if } \lambda_i = 1 \pmod 4, \\ +1 & \text{if } \lambda_i = 3 \pmod 4, \end{cases}$$

for $i = 1, \cdots, n$. Then take a permutation $\pi = (\pi(0), \pi(1), \cdots, \pi(n))$ of the elements in N_0 and a sign vector $s = (s_1, s_2, \cdots, s_n)^\top$ such that

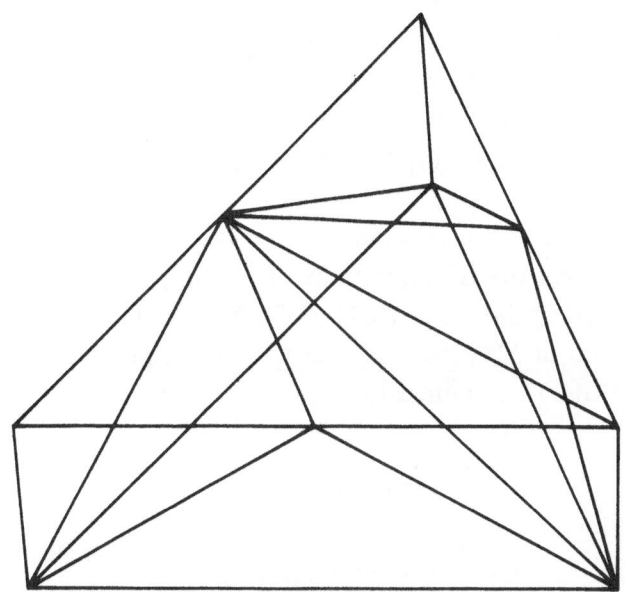

Figure 3.11: The J_3'-triangulation for $n = 2$ and $m = 1$.

$s_i \geq s_j$ if both $\lambda_i = \lambda_j$ and $i < j$, $s_{\pi(i)} = t_{\pi(i)}$ for $i = g + 1, \cdots, n$, and such that when $\lambda_{\pi(i)} = \lambda_{\pi(j)}$, if both $s_{\pi(j)} = +1$ and $i < j$ then $\pi(i) < \pi(j)$ and if both $s_{\pi(i)} = -1$ and $i < j$ then $\pi(i) > \pi(j)$, where g is the integer such that $\pi(g) = 0$.

Definition 3.3.2. For y, π and s given as above, the vectors y^{-1}, y^0, $\cdots$, y^n are given as follows:

$$y^{-1} = y,$$
$$y^i = y^{i-1} + y_0 s_{\pi(i)} h^{\pi(i)} / m, \quad i = 0, \cdots, g - 1,$$
$$y^g = y^{g-1} - y_0 \sum_{j=g+1}^{n} s_{\pi(j)} h^{\pi(j)} / m + y_0 h^0,$$
$$y^i = y^{i-1} + 2 y_0 s_{\pi(i)} h^{\pi(i)} / m, \quad i = g + 1, \cdots, n.$$

Let J_3' denote the collection of simplices $J_3'(y, \pi, s)$ that are the convex hull of y^{-1}, y^0, $\cdots$, y^n, as obtained from **Definition 3.3.2**, for all y, π and s given as above. Then J_3' forms a simplicial subdivision of continuous refinement of grid sizes of $(0, 1] \times S^n$, called the J_3'-triangulation. It is illustrated in Figure 3.11 for $n = 2$ and $m = 1$.

Finally, a triangulation is given with arbitrary refinement of grid sizes of $(0,1] \times S^n$ as proposed by van der Laan and Talman in [111]. Let N_1 denote the index set $\{1, 2, \cdots, n+1\}$. Take a sequence of integers $d_0, d_1, \cdots$, such that $d_0 = m > 0$ and for $i = 0, 1, \cdots, d_{i+1} = k_i d_i$, where k_i is an arbitrary positive integer. First of all, for $i = 0, 1, \cdots$, we subdivide $[d_0/d_{i+1}, d_0/d_i] \times S^n$ into simplices such that all grid points are contained in $\{d_0/d_i\} \times S^n \cup \{d_0/d_{i+1}\} \times S^n$ and on level d_0/d_{i+1} the triangulation of S^n is a refinement with factor k_i of the one on level d_0/d_i. Combining the triangulations of $[d_0/d_{i+1}, d_0/d_i] \times S^n$ over all i, we obtain a triangulation of $(0,1] \times S^n$. For $i = 0, 1, \cdots$, let G_i be the Q-triangulation of S^n with grid size d_i^{-1}. For a grid point w of G_i, let the n-vector $\alpha(w)$ be defined by

$$\alpha_j(w) = (1 - \sum_{l=1}^{j} w_l)d_i$$

for $j = 1, 2, \cdots, n$, and the number $c(w)$ by

$$c(w) = 1 + (\sum_{j=1}^{n} \alpha_j(w)) \bmod (n+1).$$

For $i = 1, 2, \cdots, n+1$, let q^i be the i-th column of the $(n+1) \times (n+1)$-matrix Q defined by

$$Q = \begin{bmatrix} -1 & 0 & \cdots & 0 & +1 \\ +1 & -1 & \cdots & 0 & 0 \\ \vdots & \vdots & \ddots & \vdots & \vdots \\ 0 & 0 & \cdots & -1 & 0 \\ 0 & 0 & \cdots & +1 & -1 \end{bmatrix}.$$

For a simplex $\sigma = Q(y, \pi) \in G_i$, we give a new representation of this simplex by $\sigma = Q(r, \rho)$ with vertices $r^1, r^2, \cdots, r^{n+1}$ such that $c(r^i) = i$ for $i = 1, 2, \cdots, n+1$, where r is a grid point and $\rho = (\rho(1), \rho(2), \cdots, \rho(n+1))$ is a permutation of the elements in N_1 satisfying that

$$r^1 = r,$$
$$r^j = r^{j-1} + q^{\rho(j)}/d_i, \quad j = 2, 3, \cdots, n+1.$$

Next, let p_j^i, $j = 1, 2, \cdots, n+1$, be nonnegative integers with sum equal to k_i. For a given simplex $\sigma = Q(r, \rho) \in G_i$ with vertices $r^1, r^2, \cdots$,

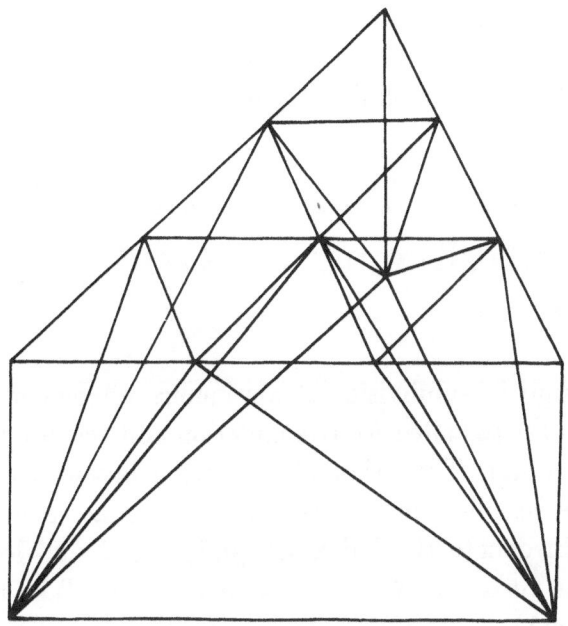

Figure 3.12: The triangulation with arbitrary refinement for $n = 2$.

r^{n+1}, we define the center point $v(\sigma)$ of σ by

$$v(\sigma) = \sum_{j=1}^{n+1} p_j^i r^j / k_i.$$

Note that $v(\sigma)$ is a grid point of G_{i+1}. Let T be a proper subset of N_1. Then we define the finite subset $A(T, \sigma)$ of grid points in G_{i+1} by

$$A(T, \sigma) = \left\{ x \in \sigma \ \middle| \ \begin{array}{l} x = v(\sigma) + \sum_{j \in T} \mu_j q^j / d_{i+1} \\ \text{for positive integers } \mu_j, \ j \in T \end{array} \right\}.$$

Now a triangulation of $[d_0/d_{i+1}, d_0/d_i] \times \sigma$ is obtained by connecting all the grid points in $A(T, \sigma)$ on level $\{d_0/d_{i+1}\} \times S^n$ with the vertices r^i, $i \notin T$, on level $\{d_0/d_i\} \times S^n$ for all T. The union of such triangulations over all $\sigma \in G_i$ forms a triangulation of $[d_0/d_{i+1}, d_0/d_i] \times S^n$. Therefore, we obtain a triangulation with arbitray refinement of grid sizes of $(0, 1] \times S^n$ by combining the triangulations of $[d_0/d_{i+1}, d_0/d_i] \times S^n$ over all i. This triangulation is illustrated in Figure 3.12 for $n = 2$.

We remark that other triangulations with arbitrary refinement of grid sizes of $(0, 1] \times S^n$ have been given by Shamir in [166]. Shamir's triangulations are quite similar to the above triangulation.

3.4   Existing Triangulations of Continuous Refinement of Grid Sizes of $(0, 1] \times R^n$

The first simplicial subdivision of continuous refinement of grid sizes of $(0, 1] \times R^n$ is the so called K_3-triangulation and was proposed by Eaves and Saigal in [42]. This triangulation has a factor of grid refinement of two. The simplices of the K_3-triangulation are defined as follows. Let N_0 again denote the index set $\{0, 1, \cdots, n\}$ and let u^i represent the i-th unit vector in R^{n+1} for $i = 0, 1, \cdots, n$. Take a permutation $\pi = (\pi(0), \pi(1), \cdots, \pi(n))$ of the elements in N_0. Let g denote the integer such that $\pi(g) = 0$. For an integer $k \geq 1$, choose a vector y such that $y_0 = 2^{-k}$, $y_{\pi(i)}/y_0$ is an integer for $i = 0, \cdots, g - 1$, and $y_{\pi(i)}/y_0$ is odd for $i = g + 1, \cdots, n$. Then define

$$w_{\pi(i)} = \begin{cases} 0 & \text{if } y_{\pi(i)}/y_0 \text{ is odd,} \\ 1 & \text{otherwise} \end{cases}$$

for $i = 0, 1, \cdots, g - 1$.

Definition 3.4.1. For y and π given as above, the vectors y^{-1}, y^0, $\cdots$, y^n are given as follows:

$$y^{-1} = y,$$
$$y^i = y^{i-1} + y_0 u^{\pi(i)}, \ i = 0, \cdots, g - 1,$$
$$y^g = y^{g-1} - y_0(\textstyle\sum_{j=0}^{g-1} w_{\pi(j)} u^{\pi(j)} + \sum_{j=g+1}^n u^{\pi(j)}) + y_0 u^0,$$
$$y^i = y^{i-1} + 2y_0 u^{\pi(i)}, \ i = g + 1, \cdots, n.$$

Let K_3 denote the collection of simplices $K_3(y, \pi)$ that are the convex hull of y^{-1}, y^0, $\cdots$, y^n, as obtained from **Definition 3.4.1**, for all y and π given as above. Then K_3 is a simplicial subdivision of continuous refinement of grid sizes of $(0, 1] \times R^n$, called the K_3-triangulation. It is illustrated in Figure 3.13 for $n = 2$.

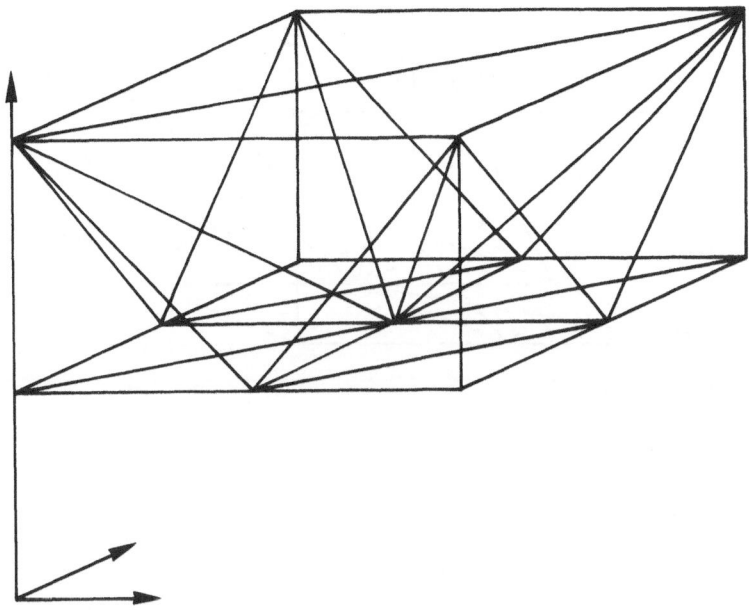

Figure 3.13: The K_3-triangulation for $n = 2$.

Next we introduce the J_3-triangulation of continuous refinement of grid sizes of $(0,1] \times R^n$ proposed by Todd in [177]. According to numerical experience this triangulation is more efficient than the K_3-triangulation. Take a permutation $\pi = (\pi(0), \pi(1), \cdots, \pi(n))$ of the elements in N_0. Let g denote the integer such that $\pi(g) = 0$. For an integer $k \geq 1$, choose a vector y such that $y_0 = 2^{-k}$ and y_i/y_0 is odd for all i. Let us define

$$t_i = \begin{cases} -1 & \text{if } y_i/y_0 = 1 (\text{mod} 4), \\ +1 & \text{if } y_i/y_0 = 3 (\text{mod} 4), \end{cases}$$

for $i = 1, 2, \cdots, n$. Choose a sign vector $s = (s_1, s_2, \cdots, s_n)^\mathsf{T}$ such that $s_{\pi(i)} \in \{-1, +1\}$ for $i = 0, \cdots, g-1$, and $s_{\pi(i)} = t_{\pi(i)}$ for $i = g+1, \cdots, n$.

Definition 3.4.2. For y, π and s given as above, the vectors y^{-1}, y^0,

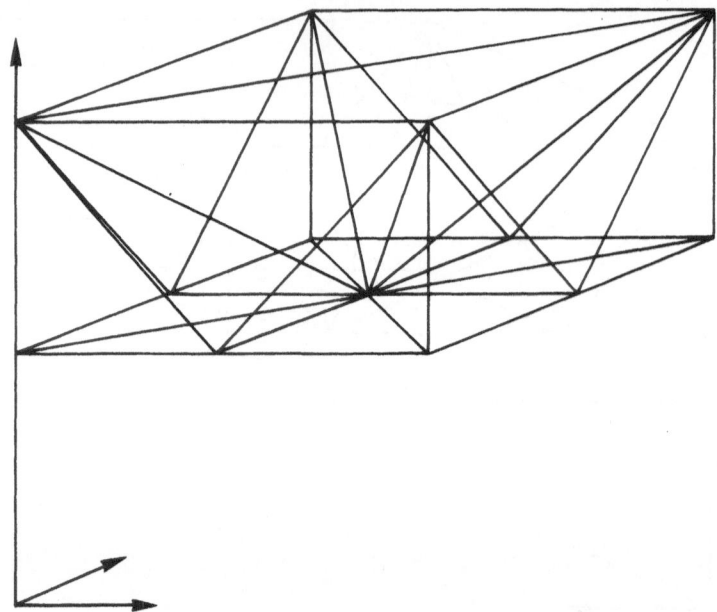

Figure 3.14: The J_3-triangulation for $n = 2$.

$\cdots, y^n$ are given as follows:

$$y^{-1} = y,$$
$$y^i = y^{i-1} + y_0 s_{\pi(i)} u^{\pi(i)}, \ i = 0, \cdots, g-1,$$
$$y^g = y^{g-1} - y_0 \sum_{j=g+1}^n s_{\pi(j)} u^{\pi(j)} + y_0 u^0$$
$$y^i = y^{i-1} + 2y_0 s_{\pi(i)} u^{\pi(i)}, \ i = g+1, \cdots, n.$$

Let J_3 denote the collection of simplices $J_3(y, \pi, s)$ that are the convex hull of $y^{-1}, y^0, \cdots, y^n$, as obtained from **Definition 3.4.2**, for all y, π and s given as above. Then J_3 forms a simplicial subdivision of continuous refinement of grid sizes of $(0, 1] \times R^n$, called the J_3-triangulation. Also this triangulation has a factor of grid refinement of two. It is illustrated in Figure 3.14 for $n = 2$.

One can use the same approach as that in the previous section to obtain a triangulation of continuous arbitrary refinement of grid sizes of $(0, 1] \times R^n$. This triangulation is illustrated in Figure 3.15 for $n = 2$.

We remark that other triangulations with arbitrary refinement of grid sizes of $(0, 1] \times R^n$ have bee proposed by Shamir in [166], Ko-

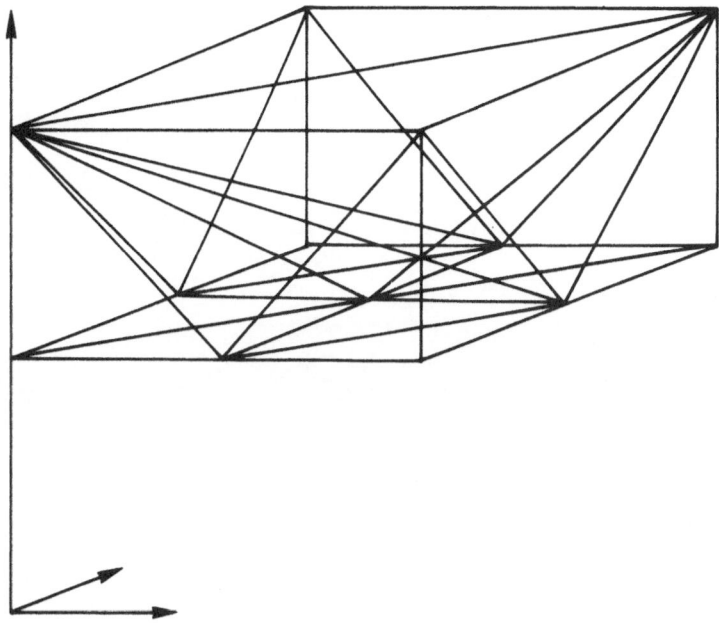

Figure 3.15: The triangulation with arbitrary refinement for $n = 2$.

jima and Yamamoto in [95], and Broadie and Eaves in [5]. Especially, Broadie and Eaves have considered how to combine simplicial homotopy algorithms with simplicial variable dimension algorithms.

Chapter 4

The D_1-Triangulation of R^n

The efficiency of simplicial algorithms depends heavily on the underlying triangulation and every different simplicial algorithm needs a triangulation suitable to itself. Therefore, in order to develop more efficient simplicial algorithms, it is important to introduce triangulations that are superior to other triangulations according to certain measures. In this chapter, we introduce a new triangulation of R^n. It is called the D_1-triangulation. This simplicial subdivision of R^n is quite different in structure from the triangulations of R^n discussed in the previous chapter. It has the important property that it subdivides every unit cube into simplices and that it can directly be applied to the Sandwich method and both the 2-ray and the $2n$-ray variable dimension methods. The D_1-triangulation is superior to both the K_1-triangulation and the J_1-triangulation according to measures of efficiency such as the number of simplices in a unit cube, the diameter, and the average directional density. This chapter is organized as follows. In Section 1, the D_1-triangulation is given. Its pivot rules are described in Section 2. We compare the D_1-triangulation with several other triangulations of R^n according to three measures of efficiency in Section 3, 4, and 5, respectively. This chapter is based on Dang's [13].

4.1   The D_1-Triangulation of R^n

For ease of notation, again let N denote the index set $\{1, 2, \cdots, n\}$, let N_0 denote the index set $\{0, 1, \cdots, n\}$, and let u^i denote the i-th unit vector in R^n for $i = 1, 2, \cdots, n$. Assume $n \geq 2$. The simplices of the D_1-triangulation of R^n are defined as follows.

Let D denote the set $\{y \in R^n \mid \text{all components of } y \text{ are even}\}$. Take $y \in D$. Then a sign vector s in R^n is chosen such that $s_i \in \{-1, +1\}$ for all $i \in N$ and an integer p such that $0 \leq p \leq n - 1$. Finally, take a permutation $\pi = (\pi(1), \pi(2), \cdots, \pi(n))$ of the n elements of N.

Definition 4.1.1. For y, π, s, and p given as above, the vectors y^0, y^1, $\cdots$, y^n are given as follows. If $p = 0$, then $y^0 = y$ and

$$y^k = y + s_{\pi(k)} u^{\pi(k)}, \ k = 1, 2, \cdots, n.$$

If $p \geq 1$, then $y^0 = y + s$,

$$y^k = y^{k-1} - s_{\pi(k)} u^{\pi(k)}, \ k = 1, 2, \cdots, p - 1, \text{ and}$$
$$y^k = y + s_{\pi(k)} u^{\pi(k)}, \ k = p, \cdots, n.$$

Lemma 4.1.2. Let y^0, y^1, $\cdots$, y^n be obtained from **Definition 4.1.1**. Then y^0, y^1, $\cdots$, y^n are affinely independent.

Proof. If $p = 0$, then let

$$
\begin{aligned}
z^1 &= y^1 - y^0 = s_{\pi(1)} u^{\pi(1)}, \\
z^2 &= y^2 - y^1 = s_{\pi(2)} u^{\pi(2)} - s_{\pi(1)} u^{\pi(1)}, \\
&\cdots, \\
z^n &= y^n - y^{n-1} = s_{\pi(n)} u^{\pi(n)} - s_{\pi(n-1)} u^{\pi(n-1)}.
\end{aligned}
$$

Obviously, z^1, z^2, $\cdots$, z^n are linearly independent.

If $p \geq 1$, then let

$$
\begin{aligned}
z^k &= y^k - y^{k-1} = -s_{\pi(k)} u^{\pi(k)}, \ k = 1, 2, \cdots, p - 1, \\
z^p &= y^p - y^{p-1} = -\sum_{k=p+1}^{n} s_{\pi(k)} u^{\pi(k)}, \\
z^k &= y^k - y^{k-1} = s_{\pi(k)} u^{\pi(k)} - s_{\pi(k-1)} u^{\pi(k-1)}, \ k = p+1, \cdots, n.
\end{aligned}
$$

Suppose that $z^1, z^2, \cdots, z^n$ are linearly dependent. Then there exists a vector $q = (q_1, q_2, \cdots, q_n)^T \neq 0$ such that

$$q_1 z^1 + q_2 z^2 + \cdots + q_n z^n = 0.$$

When $p = n - 1$, it is necessary that

$$
\begin{aligned}
q_1 &= \cdots = q_{n-2} = 0, \\
-q_{n-1} + q_n &= 0, \\
q_n &= 0.
\end{aligned}
$$

We conclude that

$$q_1 = q_2 = \cdots = q_n = 0.$$

When $1 \leq p < n - 1$, we must have that

$$
\begin{aligned}
q_1 &= q_2 = \cdots = q_{p-1} = 0, \\
q_{p+1} &= 0, \\
q_k - q_{k+1} - q_p &= 0 \text{ for } k = p + 1, \cdots, n - 1, \\
q_n - q_p &= 0.
\end{aligned}
$$

Therefore,

$$
\begin{aligned}
q_p &= q_n, \\
q_{n-1} &= 2q_n, \\
q_{n-2} &= 3q_n, \\
&\cdots, \\
q_{p+2} &= (n - (p + 1))q_n, \\
q_{p+2} + q_p &= 0.
\end{aligned}
$$

Hence,

$$(n - (p + 1) + 1)q_n = 0.$$

Since $p < n - 1$, we conclude

$$q_1 = q_2 = \cdots = q_n = 0.$$

This means that the hypothesis is incorrect, i.e., $z^1, z^2, \cdots, z^n$ are linearly independent. Therefore, $y^0, y^1, \cdots, y^n$ are affinely independent.
END

Let $y^0, y^1, \cdots, y^n$ be obtained from **Definition 4.1.1**. Then by **Lemma 4.1.2** their convex hull is an n-simplex. It is denoted by

$D_1(y, \pi, s, p)$. Let D_1 denote the collection of simplices $D_1(y, \pi, s, p)$ for all y, π, and s given as above.

Lemma 4.1.3. The union of all $\sigma \in D_1$ is equal to R^n.

Proof. Let x be an arbitrary point in R^n. For each $i \in N$, let

$$y_i = \begin{cases} \lfloor x_i \rfloor & \text{if } \lfloor x_i \rfloor \text{ is even,} \\ \lfloor x_i \rfloor + 1 & \text{otherwise,} \end{cases}$$

and

$$s_i = \begin{cases} 1 & \text{if } \lfloor x_i \rfloor \text{ is even,} \\ -1 & \text{otherwise.} \end{cases}$$

It is obvious that for all $i \in N$

$$0 \le s_i(x_i - y_i) \le 1.$$

Choose a permutation $\pi = (\pi(1), \pi(2), \cdots, \pi(n))$ of the elements of N such that

$$0 \le s_{\pi(1)}(x_{\pi(1)} - y_{\pi(1)}) \le \cdots \le s_{\pi(n)}(x_{\pi(n)} - y_{\pi(n)}) \le 1.$$

If $\sum_{i=1}^{n} s_i(x_i - y_i) \le 1$, let

$$q_1 = s_{\pi(1)}(x_{\pi(1)} - y_{\pi(1)}), \cdots, q_n = s_{\pi(n)}(x_{\pi(n)} - y_{\pi(n)}),$$

and $q_0 = 1 - \sum_{j=1}^{n} q_j$. Obviously, $q_j \ge 0$ for all j and $\sum_{j=0}^{n} q_j = 1$. Take $p = 0$. Let $y^0 = y$ and

$$y^k = y + s_{\pi(k)} u^{\pi(k)} \text{ for } k = 1, 2, \cdots, n.$$

It is easily seen that

$$x = \sum_{j=0}^{n} q_j y^j.$$

Thus $x \in D_1(y, \pi, s, p)$.

Suppose $\sum_{i=1}^{n} s_i(x_i - y_i) > 1$. We show that there exists an integer $1 \le p \le n-1$ such that the following system has a nonnegative solution,

$$\sum_{i=0}^{j-1} q_i = s_{\pi(j)}(x_{\pi(j)} - y_{\pi(j)}), \ j = 1, 2 \cdots, p - 1,$$
$$\sum_{i=0}^{p-1} q_i + q_k = s_{\pi(k)}(x_{\pi(k)} - y_{\pi(k)}), \ k = p, \cdots, n,$$
$$q_0 + q_1 + \cdots + q_n = 1.$$

In fact, rewriting the system, we obtain that

$$
\begin{aligned}
q_0 &= s_{\pi(1)}\big(x_{\pi(1)} - y_{\pi(1)}\big), \\
q_{j-1} &= s_{\pi(j)}\big(x_{\pi(j)} - y_{\pi(j)}\big) - s_{\pi(j-1)}\big(x_{\pi(j-1)} - y_{\pi(j-1)}\big), \\
& \quad j = 2, \cdots, p-1, \\
q_{p-1} &= -s_{\pi(p-1)}\big(x_{\pi(p-1)} - y_{\pi(p-1)}\big) \\
& \quad + \big(\textstyle\sum_{j=p}^{n} s_{\pi(j)}\big(x_{\pi(j)} - y_{\pi(j)}\big) - 1\big)/(n-p), \\
q_k &= s_{\pi(k)}\big(x_{\pi(k)} - y_{\pi(k)}\big) \\
& \quad + \big(1 - \textstyle\sum_{j=p}^{n} s_{\pi(j)}\big(x_{\pi(j)} - y_{\pi(j)}\big)\big)/(n-p), \\
& \quad k = p, \cdots, n.
\end{aligned}
$$

For $p = n-1$, if $q_{n-2} \geq 0$, then it is clear that $q_j \geq 0$ for all $j \in N_0$. Otherwise, there exists an integer p_0 with $1 \leq p_0 \leq n-2$ such that

$$
\begin{aligned}
0 \leq & -s_{\pi(p_0-1)}\big(x_{\pi(p_0-1)} - y_{\pi(p_0-1)}\big) \\
& + \big(\textstyle\sum_{j=p_0}^{n} s_{\pi(j)}\big(x_{\pi(j)} - y_{\pi(j)}\big) - 1\big)/(n - p_0)
\end{aligned}
$$

and

$$
\begin{aligned}
0 > & -s_{\pi(p_0)}\big(x_{\pi(p_0)} - y_{\pi(p_0)}\big) \\
& + \big(\textstyle\sum_{j=p_0+1}^{n} s_{\pi(j)}\big(x_{\pi(j)} - y_{\pi(j)}\big) - 1\big)/(n - p_0 - 1),
\end{aligned}
$$

and hence,

$$
\begin{aligned}
& s_{\pi(p_0)}\big(x_{\pi(p_0)} - y_{\pi(p_0)}\big) + \big(1 - \textstyle\sum_{j=p_0}^{n} s_{\pi(j)}\big(x_{\pi(j)} - y_{\pi(j)}\big)\big)/(n - p_0) \\
& \geq s_{\pi(p_0)}\big(x_{\pi(p_0)} - y_{\pi(p_0)}\big) + \big(1 - s_{\pi(p_0)}\big(x_{\pi(p_0)} - y_{\pi(p_0)}\big) \\
& \quad - (n - p_0 - 1)s_{\pi(p_0)}\big(x_{\pi(p_0)} - y_{\pi(p_0)}\big) - 1\big)/(n - p_0) = 0.
\end{aligned}
$$

Therefore, by taking p to be equal to p_0, $q_j \geq 0$ for all $j \in N_0$. Let $1 \leq p \leq n-1$ be the integer such that the above system has a nonnegative solution. Let $y^0 = y + s$,

$$
\begin{aligned}
y^k &= y^{k-1} - s_{\pi(k)}u^{\pi(k)}, \quad k = 1, 2, \cdots, p-1, \\
y^k &= y + s_{\pi(k)}u^{\pi(k)}, \quad k = p, \cdots, n.
\end{aligned}
$$

Then it is easily seen that

$$
x = \sum_{j=0}^{n} q_j y^j.
$$

Thus $x \in D_1(y, \pi, s, p)$.

From these results, the lemma follows immediately.

END

Lemma 4.1.4. For any σ^1 and σ^2 in D_1, $\sigma^1 \cap \sigma^2$ is either empty or a common face of both σ^1 and σ^2.

Proof. Let $x \in R^n$ be arbitrary. From **Lemma 4.1.3**, it follows that $x \in \sigma$ for some $\sigma = D_1(y, \pi, s, p)$ with vertices y^0, y^1, $\cdots$, y^n, i.e., $x = \sum_{i=0}^{n} q_i y^i$ for some $q_i \geq 0$ for all i with $\sum_{i=0}^{n} q_i = 1$. Then x lies in a face of σ whose vertices are equal to y^j for $j \in J$, where J is defined by the set $\{i \in N_0 \mid q_i > 0\}$. It is sufficient to show how y^j, $j \in J$, can be generated from x independent of y, π, s and p. Thus these vertices are found for any simplex of D_1 containing x.

For each $i \in N$, let

$$r_i = \begin{cases} \lfloor x_i \rfloor & \text{if } \lfloor x_i \rfloor \text{ is even,} \\ \lfloor x_i \rfloor + 1 & \text{otherwise,} \end{cases}$$

and

$$t_i = \begin{cases} 1 & \text{if } x_i - r_i > 0, \\ 0 & \text{if } x_i - r_i = 0, \\ -1 & \text{if } x_i - r_i < 0. \end{cases}$$

Let $w = \sum_{i=1}^{n} t_i(x_i - r_i)$. Further, let

$$y_i(t_j) = \begin{cases} r_i + t_j & \text{if } i = j, \\ r_i & \text{otherwise,} \end{cases}$$

for $i = 1, 2, \cdots, n$, and define

$$y(t_j) = (y_1(t_j), y_2(t_j), \cdots, y_n(t_j))^{\mathsf{T}}.$$

Then

$$\{y(t_1), y(t_2), \cdots, y(t_n), r\} = \left\{ y^j \mid j \in J \right\}$$

if $w < 1$, and

$$\{y(t_1), y(t_2), \cdots, y(t_n)\} \setminus r = \left\{ y^j \mid j \in J \right\}$$

if $w = 1$.

Suppose that $w > 1$. Let $T_1, T_2, \cdots, T_g$ be subsets of N such that $\cup_{k=1}^{g} T_k = N$ and that for each $1 \le k \le g$,

$$t_i(x_i - r_i) = t_j(x_j - r_j)$$

if $i, j \in T_k$ and that for any $1 \le e < f \le g$,

$$t_i(x_i - r_i) < t_j(x_j - r_j)$$

if $i \in T_e$ and $j \in T_f$. Let T_0 denote the empty set. Next let $i(k)$ denote an arbitrary element of T_k for $k = 0, 1, \cdots, g$. Since $w > 1$, there exist unique $0 \le v < g$ and $q \ge 0$ such that

$$\begin{aligned}
1 = t_{i(v)}(x_{i(v)} - r_{i(v)}) + (1 - \mid T_{v+1} \mid - \cdots - \mid T_g \mid)q \\
+ \mid T_{v+1} \mid (t_{i(v+1)}(x_{i(v+1)} - r_{i(v+1)}) - t_{i(v)}(x_{i(v)} - r_{i(v)})) + \cdots \\
+ \mid T_g \mid (t_{i(g)}(x_{i(g)} - r_{i(g)}) - t_{i(v)}(x_{i(v)} - r_{i(v)}))
\end{aligned}$$

and

$$t_{i(v+1)}(x_{i(v+1)} - r_{i(v+1)}) - t_{i(v)}(x_{i(v)} - r_{i(v)}) - q \ge 0.$$

For $0 \le k \le v$, let

$$y_i(T_k) = \begin{cases} r_i + t_i & \text{if } i \notin \cup_{l=0}^{k} T_l, \\ r_i & \text{otherwise,} \end{cases}$$

for $i = 1, 2, \cdots, n$, and define

$$y(T_k) = (y_1(T_k), y_2(T_k), \cdots, y_n(T_k))^\mathsf{T}.$$

For $v + 1 \le k \le g$, let, for each $j \in T_k$,

$$\hat{y}_i(j) = \begin{cases} r_i + t_j & \text{if } i = j, \\ r_i & \text{otherwise,} \end{cases}$$

for $i = 1, 2, \cdots, n$, and define

$$\hat{y}(j) = (\hat{y}_1(j), \hat{y}_2(j), \cdots, \hat{y}_n(j))^\mathsf{T}.$$

Furthermore, let

$$\bar{g} = \begin{cases} g - 1 & \text{if } t_{i(g)}(x_{i(g)} - r_{i(g)}) - t_{i(v)}(x_{i(v)} - r_{i(v)}) - q = 0, \\ g & \text{otherwise.} \end{cases}$$

If $q = 0$, then

$$\{y(T_k) \mid 0 \le k < v\} \cup (\cup_{k=v+1}^{\bar{g}} \{\hat{y}(j) \mid j \in T_k\}) = \{y^j \mid j \in J\},$$

and if $q > 0$, then

$$\{y(T_k) \mid 0 \le k \le v\} \cup (\cup_{k=v+1}^{\bar{g}} \{\hat{y}(j) \mid j \in T_k\}) = \{y^j \mid j \in J\}.$$

From the above results, the lemma follows immediately.
END

Theorem 4.1.5. The set D_1 is a triangulation of R^n.

Proof. Let $x \in R^n$ be arbitrary. It is clear that x is only contained in a finite number of simplices of D_1. Using **Lemma 4.1.2**, **Lemma 4.1.3**, and **Lemma 4.1.4**, we conclude the result of the theorem.
END

This simplicial subdivision of R^n is called the D_1-triangulation. In Figure 4.1, the D_1-triangulation of R^n is illustrated for $n = 2$ and $n = 3$.

4.2 Pivot Rules of the D_1-Triangulation

Let $\sigma = D_1(y, \pi, s, p)$ with vertices $y^0, y^1, \cdots, y^n$ be given. We wish to obtain the unique n-simplex, $\bar{\sigma} = D_1(\bar{y}, \bar{\pi}, \bar{s}, \bar{p})$, with vertices $\bar{y}^0, \bar{y}^1, \cdots, \bar{y}^n$ containing all the vertices of σ except vertex y^i. **Table 4.2.1** shows how $\bar{y}, \bar{\pi}, \bar{s}$, and $\bar{p}$ depend on y, π, s, p, and i. From this table it is easy to obtain each vertex $\bar{y}^k$ of $\bar{\sigma}$ for $k = 0, 1, \cdots, n$, and in particular the vertex of $\bar{\sigma}$ not being a vertex of σ.

We remark that simplices of the D_1-triangulation can be represented in more than one way. More precisely, if p is equal to 0 or 1 then for a given y and s every permutation π yields the same simplex. If $p > 1$, then for a given y and s, every permutation π with the same $\pi(1), \pi(2), \cdots, \pi(p-1)$ also yields the same simplex.

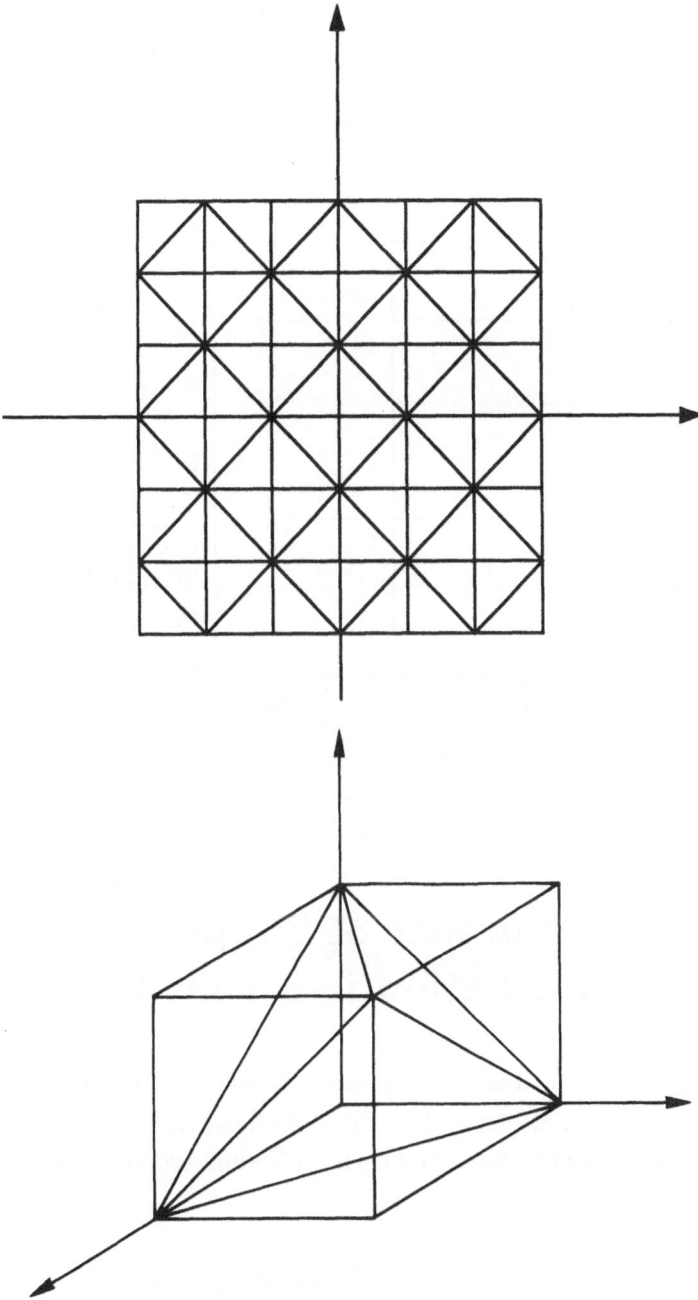

Figure 4.1: The D_1-triangulation for $n = 2$ and $n = 3$.

Table 4.2.1. The Pivot Rules of the D_1-Triangulation

i	p	$\bar{y}$	$\bar{s}$	$\bar{\pi}$	$\bar{p}$
0	0	y	s	π	1
$i \geq 1$	0	y	$s - 2s_{\pi(i)}u^{\pi(i)}$	π	p
0	1	y	s	π	0
0	$2 \leq p$	y	$s - 2s_{\pi(1)}u^{\pi(1)}$	π	p
$1 \leq i < p-1$		y	s	$(\pi(1),\cdots,\pi(i+1),\pi(i),\cdots,\pi(n))$	p
$p-1$	$2 \leq p$	y	s	π	$p-1$
$i > p-1$	$1 \leq p < n-1$	y	s	$(\pi(1),\cdots,\pi(p-1),\pi(i),\pi(p),\cdots,\pi(i-1),\pi(i+1),\cdots,\pi(n))$	$p+1$
$n-1$	$n-1$	$y + 2s_{\pi(n)}u^{\pi(n)}$	$s - 2s_{\pi(n)}u^{\pi(n)}$	π	p
n	$n-1$	$y + 2s_{\pi(n-1)}u^{\pi(n-1)}$	$s - 2s_{\pi(n-1)}u^{\pi(n-1)}$	π	p

4.3 The Number of Simplices of the D_1-Triangulation in a Unit Cube

In this and the next two sections we compare the D_1-triangulation with the other triangulations of R^n. First of all, in this section we count the number of simplices into which the different triangulations subdivide the unit cube.

Theorem 4.3.1. The number of simplices of the D_1-triangulation in the unit cube is equal to d_n defined by

$$d_n = n + n(n-1) + \cdots + n(n-1)\cdots 4 \cdot 3 + 2.$$

Proof. Let Q denote the collection of simplices in D_1 that triangulate the unit cube U^n, i.e.,

$$Q = \left\{ D_1(y, \pi, s, p) \mid y = 0, s = (1, 1, \ldots, 1)^\mathsf{T} \right\}.$$

From **Definition 4.1.2**, it follows that in Q there are only one simplex with p equal to zero, one simplex with p equal to one, and $n!/(n-q+1)!$ simplices with p equal to q, $2 \le q \le n-1$. Thus

$$
\begin{aligned}
|Q| &= 1 + 1 + n!/(n-1)! + n!/(n-2)! + \cdots + n!/2! \\
&= 2 + n + n(n-1) + \cdots + n(n-1) \cdots 4 \cdot 3.
\end{aligned}
$$

Since the union of all simplices in Q is equal to U^n, the theorem follows immediately.

END

Theorem 4.3.2. The number of simplices in the unit cube of the K_1-triangulation, the J_1-triangulation, and the H_1-triangulation is $n!$.

Proof. The proof is trivial.

END

Therefore,

$$\mathcal{N}(K_1) = \mathcal{N}(J_1) = \mathcal{N}(H_1) = n!$$

and

$$\mathcal{N}(D_1) = n + n(n-1) + \cdots + n(n-1) \cdots 4 \cdot 3 + 2.$$

Theorem 4.3.3. If $n \ge 3$, then $d_n < n!$. As n goes to infinity, $d_n/n!$ converges to $e - 2$.

Proof. For $n = 3$, we have $d_3 < 3!$ since $d_3 = 5$ and $3! = 6$. Suppose that $d_{n-1} < (n-1)!$ for $n \ge 4$. Thus $nd_{n-1} < n!$. From

$$nd_{n-1} = n(n-1) + n(n-1)(n-2) + \cdots + n(n-1) \cdots 4 \cdot 3 + 2n = d_n + (n-2),$$

we obtain that $d_n = nd_{n-1} - (n-2) < n!$. By the induction principle, the conclusion that $d_n < n!$ for $n \ge 3$ follows directly.

Furthermore,

$$d_n/n! = 1/(n-1)! + 1/(n-2)! + \cdots + 1/2! + 2/n!,$$

so $d_n/n!$ converges to $e-2$ as n goes to infinity.

END

The last theorem implies that the number of simplices of the D_1-triangulation in the unit cube is less than ones of the other triangulations. Moreover, the ratio of the number of simplices in the unit cube between the D_1-triangulation and the other ones converges to $e-2$ when n goes to infinity.

4.4   The Diameter of the D_1-Triangulation

In this section we calculate the diameter of the D_1-triangulation and compare it with the diameters of the other triangulations that also subdivide the unit cube into simplices.

Theorem 4.4.1. The diameter of the D_1-triangulation is equal to

$$\mathcal{D}(D_1) = 2n - 3.$$

Proof. Let σ be the simplex $D_1(y, \pi, s, p)$ such that $y = 0$, $s = (1, 1, \cdots, 1)^\mathsf{T}$, $p = n - 1$, and $\pi = (1, 2, \cdots, n)$ and let τ be a facet of σ opposite to the vertex $y^0 = y + s$. Clearly, τ lies in the boundary of the unit cube U^n. Next, let $\bar{\sigma}$ be the simplex $D_1(\bar{y}, \bar{\pi}, \bar{s}, \bar{p})$ such that $\bar{y} = 0$, $\bar{s} = (1, 1, \cdots, 1)^\mathsf{T}$, $\bar{p} = n - 1$ and $\bar{\pi} = (n, n-1, \cdots, 3, 2, 1)$ and let $\bar{\tau}$ be a facet of $\bar{\sigma}$ opposite to the vertex $\bar{y}^0 = \bar{y} + \bar{s}$. Then $\bar{\tau}$ also lies in the boundary of the unit cube U^n. Furthermore, let $\sigma_1, \cdots, \sigma_{m-1}$ be a sequence of simplices of D_1 in U^n such that σ_{i-1} and σ_i are adjacent for $i = 2, \cdots, m - 1$, σ and σ_1 are adjacent, and σ_{m-1} and $\bar{\sigma}$ are also adjacent. It is easily seen that the smallest m is equal to $2n - 4$. The distance between τ and $\bar{\tau}$ is obviously the greatest distance between any two facets in the boundary of U^n. Therefore,

$$\mathcal{D}(D_1) = 2n - 3.$$

END

Theorem 4.4.2. The diameters of the K_1-triangulation and of the J_1-triangulation are the same and equal to

$$\mathcal{D}(K_1) = \mathcal{D}(J_1) = 1 + n(n-1)/2,$$

and the diameter of the H_1-triangulation is such that

$$\mathcal{D}(H_1) \geq (n^3 - n + 6)/6.$$

Proof. Let σ be the simplex $K_1(y, \pi)$ such that $y = 0$ and $\pi = (1, 2, \cdots, n)$ and let τ be a facet of σ opposite to the vertex $y^n = y + u$, where u is the vector with all components equal to one. Next, let $\bar{\sigma}$ be the simplex $K_1(\bar{y}, \bar{\pi})$ such that $\bar{y} = 0$ and $\bar{\pi} = (n, n-1, \cdots, 1)$ and let $\bar{\tau}$ be a facet of $\bar{\sigma}$ opposite to the vertex $\bar{y}^n = \bar{y} + u$. Furthermore, let $\sigma_1, \cdots, \sigma_{m-1}$ be a sequence of simplices of K_1 in U^n such that σ_{i-1} and σ_i are adjacent for $i = 2, \cdots, m-1$, σ and σ_1 are adjacent, and σ_{m-1} and $\bar{\sigma}$ are also adjacent. It is easily seen that the smallest m is equal to $n(n-1)/2$. The distance between τ and $\bar{\tau}$ is obviously the greatest distance between any two facets in the boundary of U^n. Therefore,

$$\mathcal{D}(K_1) = 1 + n(n-1)/2.$$

Since the J_1-triangulation and the K_1-triangulation coincide on U^n,

$$\mathcal{D}(J_1) = \mathcal{D}(K_1).$$

Let σ be the simplex $H_1(y, \pi)$ such that $y = (1, 0, \cdots, 0)^\top$ and $\pi = (1, 2, \cdots, n)$ and let τ be a facet of σ opposite to the vertex $y^0 = y$. Next, let $\bar{\sigma}$ be the simplex $H_1(\bar{y}, \bar{\pi})$ such that $\bar{y} = (1, 1, \cdots, 1)^\top$ and $\bar{\pi} = (n, n-1, \cdots, 1)$ and let $\bar{\tau}$ be a facet of $\bar{\sigma}$ opposite to the vertex $\bar{y}^n = y - u^1$. Furthermore, let $\sigma_1, \cdots, \sigma_{m-1}$ be a sequence of simplices of H_1 in U^n such that σ_{i-1} and σ_i are adjacent for $i = 2, \cdots, m-1$, σ and σ_1 are adjacent, and σ_{m-1} and $\bar{\sigma}$ are also adjacent. It is easily seen that the smallest m is equal to $(n^3 - n + 6)/6 - 1$. Thus the distance between τ and $\bar{\tau}$ is $(n^3 - n + 6)/6$. This means that

$$\mathcal{D}(H_1) \geq (n^3 - n + 6)/6.$$

END

The above conclusions show that the diameter of the D_1-triangulation is in the order n smaller than the diameter of the K_1-triangulation or the J_1-triangulation and at least in the order n^2 smaller than the diameter of the H_1-triangulation.

4.5　The Average Directional Density of the D_1-Triangulation

From Eaves and Yorke [44], we know that for a triangulation the average directional density and the surface density are equivalent. Firstly, we calculate the surface density of the D_1-triangulation and obtain then the average directional density of the D_1-triangulation from its surface density. For the definition of the surface density we refer to Eaves and York [44].

For a simplex σ, the volume of σ is denoted by $V(\sigma)$ and the surface area of σ is denoted by $SA(\sigma)$. Let

$$\sigma^0 = \mathrm{conv}(\{0, u^1, \cdots, u^n\}),$$

$$\sigma^1 = \mathrm{conv}(\{u, u^1, u^2, \cdots, u^n\}),$$

$$\sigma^2 = \mathrm{conv}(\{u, u - u^1, u^2, \cdots, u^n\}),$$

$$\cdots,$$

$$\sigma^{n-1} = \mathrm{conv}(\{u, u - u^1, \cdots, u - u^1 - u^2 - \cdots - u^{n-2}, u^{n-1}, u^n\}).$$

The surface density of the D_1-triangulation is equal to

$$SD(D_1) = (SA(\sigma^0)/V(\sigma^0) + \sum_{k=1}^{n-1} n!/(n-k+1)! SA(\sigma^k)/V(\sigma^k))/d_n.$$

Let

$$\tau_0^0 = \mathrm{conv}(\{u^1, u^2, \cdots, u^n\}),$$

$$\tau_1^0 = \mathrm{conv}(\{0, u^2, \cdots, u^n\}),$$

$$\cdots,$$

$$\tau_n^0 = \mathrm{conv}(\{0, u^1, \cdots, u^{n-1}\}),$$

which give all of the facets of σ^0. Then the surface aera of σ^0 is equal to

$$SA(\sigma^0) = nV(\tau_n^0) + V(\tau_0^0).$$

Clearly,

$$V(\tau_n^0) = (1/(n-1)!) \, | \det[u^1, u^2, \cdots, u^n] \, | = 1/(n-1)!$$

and

$$V(\tau_0^0) = (1/(n-1)!) \, | \det[n^{-1/2}u, u^2 - u^1, \cdots, u^n - u^1] \, | = n^{1/2}/(n-1)!.$$

Therefore,

$$SA(\sigma^0) = (n + n^{1/2})/(n-1)!.$$

Since $V(\sigma^0) = 1/n!$, we obtain that

$$SA(\sigma^0)/V(\sigma^0) = n(n + n^{1/2}).$$

Let

$$\tau_0^1 = \text{conv}(\{u^1, u^2, \cdots, u^n\}),$$

$$\tau_1^1 = \text{conv}(\{u, u^2, \cdots, u^n\}),$$

$$\tau_2^1 = \text{conv}(\{u, u^1, u^3, \cdots, u^n\}),$$

$$\cdots,$$

$$\tau_n^1 = \text{conv}(\{u, u^1, u^2, \cdots, u^{n-1}\}),$$

which give all of the facets of σ^1. Then the surface aera of σ^1 is equal to

$$SA(\sigma^1) = nV(\tau_n^1) + V(\tau_0^1).$$

Let

$$q_1 = \cdots = q_{n-1} = (n^2 - 3n + 3)^{-1/2}$$

and

$$q_n = -(n-2)(n^2 - 3n + 3)^{-1/2}.$$

Then

$$V(\tau_n^1) = (1/(n-1)!) \, | \det \begin{bmatrix} 0 & 1 & \cdots & 1 & q_1 \\ 1 & 0 & \cdots & 1 & q_2 \\ \vdots & \vdots & \ddots & \vdots & \vdots \\ 1 & 1 & \cdots & 0 & q_{n-1} \\ 1 & 1 & \cdots & 1 & q_n \end{bmatrix} |$$

$$= (n^2 - 3n + 3)^{1/2}/(n-1)!.$$

Moreover,

$$V(\tau_0^1) = n^{1/2}/(n-1)! \text{ and } V(\sigma^1) = (n-1)/n!.$$

Therefore,

$$SA(\sigma^1) = (n(n^2 - 3n + 3)^{1/2} + n^{1/2})/(n-1)!.$$

Hence,

$$SA(\sigma^1)/V(\sigma^1) = n(n(n^2 - 3n + 3)^{1/2} + n^{1/2})/(n-1).$$

For $k = 2, \cdots, n-1$, let

$$\tau_0^k = \text{conv}(\{u - u^1, \cdots, u - u^1 - \cdots - u^{k-1}, u^k, \cdots, u^n\}),$$

$$\begin{aligned}\tau_j^k = \ & \text{conv}(\{u, u - u^1, \cdots, u - u^1 - \cdots - u^{j-1}, \\ & u - u^1 - \cdots - u^{j+1}, \cdots, u - u^1 - \cdots - u^{k-1}, u^k, \cdots, u^n\}),\end{aligned}$$

$$j = 1, \cdots, k-1,$$

and

$$\begin{aligned}\tau_j^k = \ & \text{conv}(\{u, u - u^1, \cdots, u - u^1 - \cdots - u^{k-1}, \\ & u^k, \cdots, u^{j-1}, u^{j+1}, \cdots, u^n\}),\end{aligned}$$

$$j = k, \cdots, n,$$

which give all of the facets of σ^k. Then the surface aera of σ^k is equal to

$$SA(\sigma^k) = V(\tau_0^k) + \sum_{j=1}^{k-1} V(\tau_j^k) + (n - k + 1)V(\tau_n^k)$$

for $k = 2, \cdots, n - 1$. Directly,

$$V(\tau_0^k) = (1/(n-1)!) \left| \det \begin{bmatrix} 0 & 0 & \cdots & 0 & 0 & \cdots & 0 & 1 \\ 1 & 1 & \cdots & 1 & 1 & \cdots & 1 & 0 \\ 0 & 1 & \cdots & 1 & 1 & \cdots & 1 & 0 \\ \vdots & \vdots & \ddots & \vdots & \vdots & & \vdots & \vdots \\ 0 & 0 & \cdots & 1 & 1 & \cdots & 1 & 0 \\ 0 & 0 & \cdots & 0 & 0 & \cdots & 1 & 0 \\ \vdots & \vdots & & \vdots & \vdots & \ddots & \vdots & \vdots \\ 0 & 0 & \cdots & 0 & 1 & \cdots & 0 & 0 \end{bmatrix} \right|$$

$$= (n - k)/(n - 1)!.$$

Now suppose that $1 \leq j \leq k - 1$. If $j < k - 1$, let

$$q_j = 2^{-1/2}, \quad q_{j+1} = -2^{-1/2},$$
$$q_{j+2} = \cdots = q_n = 0.$$

If $j = k - 1$, let

$$q_{k-1} = -(n - k)((n - k + 1)^2 - (n - k + 1) + 1)^{-1/2}$$

and

$$q_k = \cdots = q_n = ((n - k + 1)^2 - (n - k + 1) + 1)^{-1/2}.$$

Then for every $j \in \{1, 2, \cdots, k-1\}$,

$$V(\tau_j^k) = (1/(n-1)!) \left| \det \begin{bmatrix} 1 & 1 & \cdots & 1 & 1 & \cdots & 1 & 1 & \cdots & 1 & 0 \\ 0 & 1 & \cdots & 1 & 1 & \cdots & 1 & 1 & \cdots & 1 & 0 \\ \vdots & \vdots & \ddots & \vdots & \vdots & & \vdots & \vdots & & \vdots & \vdots \\ 0 & 0 & \cdots & 1 & 1 & \cdots & 1 & 1 & \cdots & 1 & 0 \\ 0 & 0 & \cdots & 0 & 1 & \cdots & 1 & 1 & \cdots & 1 & q_j \\ 0 & 0 & \cdots & 0 & 1 & \cdots & 1 & 1 & \cdots & 1 & q_{j+1} \\ \vdots & \vdots & & \vdots & \vdots & \ddots & \vdots & \vdots & & \vdots & \vdots \\ 0 & 0 & \cdots & 0 & 0 & \cdots & 1 & 1 & \cdots & 1 & q_{k-1} \\ 0 & 0 & \cdots & 0 & 0 & \cdots & 0 & 0 & \cdots & 1 & q_k \\ \vdots & \vdots & & \vdots & \vdots & & \vdots & \vdots & \ddots & \vdots & \vdots \\ 0 & 0 & \cdots & 0 & 0 & \cdots & 0 & 1 & \cdots & 0 & q_n \end{bmatrix} \right|$$

$$= \begin{cases} 2^{1/2}(n-k)/(n-1)! & \text{if } j \neq k-1, \\ ((n-k+1)^2 - (n-k+1) + 1)^{1/2}/(n-1)! & \text{if } j = k-1. \end{cases}$$

Let

$$q_1 = q_2 = \cdots = q_{n-k} = ((n-k+1)^2 - 3(n-k+1) + 3)^{-1/2}$$

and

$$q_{n-k+1} = -(n-k-1)((n-k+1)^2 - 3(n-k+1) + 3)^{-1/2}.$$

Then

$$V(\tau_n^k) = (1/(n-1)!) \left| \det \begin{bmatrix} 1 & 1 & \cdots & 1 & 1 & \cdots & 1 & 0 \\ 0 & 1 & \cdots & 1 & 1 & \cdots & 1 & 0 \\ \vdots & \vdots & \ddots & \vdots & \vdots & & \vdots & \vdots \\ 0 & 0 & \cdots & 1 & 1 & \cdots & 1 & 0 \\ 0 & 0 & \cdots & 0 & 0 & \cdots & 1 & q_1 \\ \vdots & \vdots & & \vdots & \vdots & \ddots & \vdots & \vdots \\ 0 & 0 & \cdots & 0 & 1 & \cdots & 0 & q_{n-k} \\ 0 & 0 & \cdots & 0 & 1 & \cdots & 1 & q_{n-k+1} \end{bmatrix} \right|$$

$$= ((n-k+1)^2 - 3(n-k+1) + 3)^{1/2}/(n-1)!.$$

Thus,

$$
\begin{aligned}
SA(\sigma^k) = {} & (n-k)/(n-1)! \\
& + (n-k+1)((n-k+1)^2 - 3(n-k+1)+3)^{1/2}/(n-1)! \\
& + 2^{1/2}(k-2)(n-k)/(n-1)! \\
& + ((n-k+1)^2 - (n-k+1)+1)^{1/2}/(n-1)!.
\end{aligned}
$$

Moreover,

$$
V(\sigma^k) = (1/n!)\left|\det
\begin{bmatrix}
1 & 1 & \cdots & 1 & 1 & \cdots & 1 \\
0 & 1 & \cdots & 1 & 1 & \cdots & 1 \\
\vdots & \vdots & \ddots & \vdots & \vdots & & \vdots \\
0 & 0 & \cdots & 1 & 1 & \cdots & 1 \\
0 & 0 & \cdots & 0 & 0 & \cdots & 1 \\
\vdots & \vdots & & \vdots & \vdots & \ddots & \vdots \\
0 & 0 & \cdots & 0 & 1 & \cdots & 0
\end{bmatrix}
\right|
$$

$$
= (n-k)/n!.
$$

Hence

$$
\begin{aligned}
SA(\sigma^k)/V(\sigma^k) = {} & n((n-k) + 2^{1/2}(k-2)(n-k) \\
& + (n-k+1)((n-k+1)^2 - 3(n-k+1)+3)^{1/2} \\
& + ((n-k+1)^2 - (n-k+1)+1)^{1/2})/(n-k).
\end{aligned}
$$

From the above results we obtain that the surface density of the D_1-triangulation is equal to

$$
\begin{aligned}
SD(D_1) = {} & (n(n+n^{1/2}) \\
& + \sum_{k=2}^{n-1} n!/(n-k+1)!(n((n-k) + 2^{1/2}(k-2)(n-k) \\
& + (n-k+1)((n-k+1)^2 - 3(n-k+1)+3)^{1/2} \\
& + ((n-k+1)^2 - (n-k+1)+1)^{1/2})/(n-k)) \\
& + n(n(n^2 - 3n + 3)^{1/2} + n^{1/2})/(n-1))/d_n.
\end{aligned}
$$

Let

$$
g_n = \Gamma(n/2)/((n-1)\Gamma(1/2)\Gamma((n-1)/2)).
$$

From [44] we know that the average directional density of a triangulation is g_n times its surface density. Hence, the average directional density of the D_1-triangulation is equal to

$$\mathcal{A}(D_1) = SD(D_1)g_n.$$

It is well-known that the average directional density of both the K_1-triangulation and the J_1-triangulation is equal to

$$\mathcal{A}(K_1) = \mathcal{A}(J_1) = n(2 + (n-1)2^{1/2})g_n.$$

Since for an arbitrary simplex σ_{D_1} in D_1 and an arbitrary simplex σ_{K_1} in K_1 we have that when $n > 2$,

$$SA(\sigma_{D_1})/V(\sigma_{D_1}) < SA(\sigma_{K_1})/V(\sigma_{K_1}).$$

Therefore,

$$\mathcal{A}(D_1) < \mathcal{A}(K_1) = \mathcal{A}(J_1)$$

when $n > 2$. Thus, the average directional density of the D_1-triangulation is smaller than the average directional density of the K_1-triangulation and the J_1-triangulation.

The A^*-triangulation of R^n was obtained by van der Laan and Talman in [110] from the transformation of the K_1-triangulation. Its average directional density is smaller than the average directional density of the D_1-triangulation. However, the A^*-triangulation can not induce a simplicial subdivision of a unit cube and is only suitable for one of the available variable dimension algorithms on R^n. Finally in **Table 4.5.1** we give all the results obtained above.

Table 4.5.1.

Triangulation	Number of Simplices	Diameter	Average Directional Density
$K_1(J_1)$	$n!$	$1 + n$ $(n-1)/2$	$n(2 + 2^{1/2}(n-1))g_n$
D_1	$n + n(n-1) + \cdots$ $+ n(n-1)\cdots 4 \cdot 3 + 2$	$2n - 3$	$SD(D_1)g_n$

Chapter 5

The T_1-Triangulation of the Unit Simplex

Simplicial subdivisions of the unit simplex S^n that underlie simplicial algorithms have until now only been based on the K_1-triangulation, the J_1-triangulation, or the H_1-triangulation of R^n. In the previous chapter we saw that the D_1-triangulation is superior to all these triangulations according to measures of efficiency such as the number of simplices, the diameter, and the average directional density. Therefore, it is interesting to develop simplicial subdivisions of the unit simplex S^n that are suitable to underly simplicial algorithms on S^n and being based on the D_1-triangulation of R^n. In this chapter, we propose such a simplicial subdivision, called the T_1-triangulation of S^n. It induces a suitable simplicial subdivision for the $(n + 1)$-ray variable dimension method on the unit simplex proposed by van der Laan and Talman in [108]. Section 1 introduces the new simplicial subdivision of S^n, Section 2 describes its pivot rules, and Section 3 gives a comparison on the triangulations of the unit simplex. This chapter is based on Dang and Talman's [17].

5.1 The T_1-Triangulation

Assume $n \geq 2$. Let N denote the index set $\{1, 2, \cdots, n\}$. For $i = 1, 2, \cdots, n$, let u^i be the i-th unit vector in R^n. Next, let a positive integer m be given. Further, let us define the set $C^n(m)$ by

$$C^n(m) = \{x \in R^n \mid m \ge x_1 \ge x_2 \ge \cdots \ge x_n \ge 0\}.$$

We will construct a simplicial subdivision of $C^n(m)$. Then the T_1-triangulation of the unit simplex S^n with grid size m^{-1} is obtained from this triangulation.

Let D denote the set $\{y \in C^n(m) \mid$ all components of y are even$\}$. Let $y_0 = m$ and $y_{n+1} = 0$. Take $y \in D$. Let us define

$$I(y) = \{i \in N \mid y_{i+1} < y_i < y_{i-1}\},$$

$$I^+(y) = \{i \in N \mid y_{i+1} = y_i < y_{i-1}\},$$

$$I^-(y) = \{i \in N \mid y_{i+1} < y_i = y_{i-1}\},$$

$$J(y) = \{i \in N \mid y_{i+1} = y_i = y_{i-1}\}.$$

Then it is obvious that for each $y \in D$,

$$N = I(y) \cup I^+(y) \cup I^-(y) \cup J(y).$$

Take a sign vector $s = (s_1, s_2, \cdots, s_n)^\top$ such that for $j \in N \backslash I(y)$, if $s_j = 1$ then $s_k = 1$ for all $k < j$ with $y_k = y_j$, if $s_j = -1$ then $s_k = -1$ for all $k > j$ with $y_k = y_j$, and if $y_1 = m$ then $s_1 = -1$, and if $y_n = 0$ then $s_n = 1$. Let us define the set $K(y, s)$ by

$$K(y, s) = \left\{ i \in N \,\middle|\, \begin{array}{l} i \in I(y), \text{or } i \in I^+(y) \text{ and } s_i = 1, \\ \text{or } i \in I^-(y) \text{ and } s_i = -1 \end{array} \right\}.$$

Take a permutation $\pi = (\pi(1), \pi(2), \cdots, \pi(n))$ of the elements of N such that for k and j in $N \backslash I(y)$ with both $k < j$ and $y_j = y_k$, if $s_j = 1$ then $\pi^{-1}(k) > \pi^{-1}(j)$ and if $s_k = -1$ then $\pi^{-1}(k) < \pi^{-1}(j)$. Let q denote the nonnegative integer such that $\pi(i) \in K(y, s)$ for $i = n-q+1, \cdots, n$ and $\pi(n - q) \notin K(y, s)$. Take an integer p such that $0 \le p \le q - 1$.

Definition 5.1.1. For the vector y, the permutation π, the sign vector s, and the number p given as above, the vectors $y^0, y^1, \cdots, y^n$ are given as follows.

When $q < 2$, $y^0 = y + s$ and

$$y^k = y^{k-1} - s_{\pi(k)} u^{\pi(k)}, \ \ k = 1, 2, \cdots, n.$$

When $2 \leq q < n$, $y^0 = y + s$,

$$y^k = y^{k-1} - s_{\pi(k)} u^{\pi(k)}, \ \ k = 1, 2, \cdots, n - q - 1,$$

and if $p = 0$ then $y^{n-q} = y$ and

$$y^k = y + s_{\pi(k)} u^{\pi(k)}, \ \ k = n - q + 1, \cdots, n,$$

and if $p \geq 1$ then

$$y^k = y^{k-1} - s_{\pi(k)} u^{\pi(k)}, \ \ k = n - q, \cdots, n - q + p - 1,$$
$$y^k = y + s_{\pi(k)} u^{\pi(k)}, \ \ k = n - q + p, \cdots, n.$$

When $q = n$, if $p = 0$ then $y^0 = y$ and

$$y^k = y + s_{\pi(k)} u^{\pi(k)}, \ \ k = 1, 2, \cdots, n,$$

and if $p \geq 1$ then $y^0 = y + s$,

$$y^k = y^{k-1} - s_{\pi(k)} u^{\pi(k)}, \ \ k = 1, 2, \cdots, p - 1,$$
$$y^k = y + s_{\pi(k)} u^{\pi(k)}, \ \ k = p, p + 1, \cdots, n.$$

Let y^0, y^1, $\cdots$, y^n be obtained as given in the definition. Then it is obvious that they are affinely independent. Thus their convex hull is a simplex. The vectors y^0, y^1, $\cdots$, y^n are the vertices of this simplex. Let us denote this simplex by $T_1(y, \pi, s, p)$. Further, let T_1 denote the set of simplices $T_1(y, \pi, s, p)$ for all y, π, s and p given as above. As follows, we show that T_1 is a triangulation of $C^n(m)$.

Lemma 5.1.2. The union of all simplices in T_1 is equal to $C^n(m)$.

Proof. Clearly, every simplex in T_1 is contained in $C^n(m)$. Now let $x \in C^n(m)$ be arbitrary. Then $x \in T_1(y, \pi, s, p)$ with y, π, s, and p determined as follows. The vector y is equal to

$$y_i = \begin{cases} \lfloor x_i \rfloor + 1 & \text{if } \lfloor x_i \rfloor \text{ is odd and } \lfloor x_i \rfloor < m, \\ \lfloor x_i \rfloor - 1 & \text{if } \lfloor x_i \rfloor \text{ is odd and } \lfloor x_i \rfloor = m, \\ \lfloor x_i \rfloor & \text{otherwise,} \end{cases}$$

for $i = 1, 2, \cdots, n$, and the sign vector s is equal to

$$s_i = \begin{cases} -1 & \text{if } \lfloor x_i \rfloor \text{ is odd and } \lfloor x_i \rfloor < m, \text{ or } \lfloor x_i \rfloor \text{ is even and } \lfloor x_i \rfloor = m, \\ 1 & \text{otherwise,} \end{cases}$$

for $i = 1, 2, \cdots, n$. It is obvious that $y \in D$. The permutation π is such that

$$s_{\pi(1)}(x_{\pi(1)} - y_{\pi(1)}) \le s_{\pi(2)}(x_{\pi(2)} - y_{\pi(2)}) \le \cdots \le s_{\pi(n)}(x_{\pi(n)} - y_{\pi(n)})$$

and for $i < j$ with $s_{\pi(i)} = s_{\pi(j)}$ and $y_{\pi(i)} = y_{\pi(j)}$, if $s_{\pi(i)} = -1$ then $\pi(i) < \pi(j)$ and if $s_{\pi(i)} = 1$ then $\pi(i) > \pi(j)$.

When $q < 2$, then p is equal to 0. Let

$$\begin{aligned} \beta_0 &= s_{\pi(1)}(x_{\pi(1)} - y_{\pi(1)}), \\ \beta_1 &= s_{\pi(2)}(x_{\pi(2)} - y_{\pi(2)}) - s_{\pi(1)}(x_{\pi(1)} - y_{\pi(1)}), \\ &\cdots, \\ \beta_{n-1} &= s_{\pi(n)}(x_{\pi(n)} - y_{\pi(n)}) - s_{\pi(n-1)}(x_{\pi(n-1)} - y_{\pi(n-1)}), \\ \beta_n &= 1 - s_{\pi(n)}(x_{\pi(n)} - y_{\pi(n)}). \end{aligned}$$

It is obvious that $\beta_k \ge 0$ for all k and that

$$\sum_{k=0}^{n} \beta_k = 1 \text{ and } x = \sum_{k=0}^{n} \beta_k y^k,$$

where y^k is as defined in **Definition 5.1.1** for $k = 0, 1, \cdots, n$. Thus

$$x \in T_1(y, \pi, s, p).$$

When $q = n$, the proof is the same as that of **Lemma 4.1.3**. Suppose $2 \le q < n$. Let

$$\mu = -q s_{\pi(n-q)}(x_{\pi(n-q)} - y_{\pi(n-q)}) + \sum_{k=n-q}^{n} s_{\pi(k)}(x_{\pi(k)} - y_{\pi(k)}).$$

If $\mu \le 1$, then p is equal to 0. Let

$$\begin{aligned} \beta_0 &= s_{\pi(1)}(x_{\pi(1)} - y_{\pi(1)}), \\ \beta_1 &= s_{\pi(2)}(x_{\pi(2)} - y_{\pi(2)}) - s_{\pi(1)}(x_{\pi(1)} - y_{\pi(1)}), \\ &\cdots, \\ \beta_{n-q-1} &= s_{\pi(n-q)}(x_{\pi(n-q)} - y_{\pi(n-q)}) - s_{\pi(n-q-1)}(x_{\pi(n-q-1)} - y_{\pi(n-q-1)}), \\ \beta_{n-q} &= 1 - \mu, \\ \beta_j &= s_{\pi(j)}(x_{\pi(j)} - y_{\pi(j)}) - s_{\pi(n-q)}(x_{\pi(n-q)} - y_{\pi(n-q)}), \end{aligned}$$

for $j = n - q + 1, \cdots, n$. It is obvious that $\beta_k \geq 0$ for all k and that

$$\sum_{k=0}^{n} \beta_k = 1 \text{ and } x = \sum_{k=0}^{n} \beta_k y^k,$$

where y^k is as defined in **Definition 5.1.1** for $k = 0, 1, \cdots, n$. Thus

$$x \in T_1(y, \pi, s, p).$$

Now suppose $\mu > 1$. Then p, $1 \leq p \leq q - 1$, is the integer such that the following system has a nonnegative solution,

$$
\begin{aligned}
\beta_0 &= s_{\pi(1)}(x_{\pi(1)} - y_{\pi(1)}), \\
\beta_1 &= s_{\pi(2)}(x_{\pi(2)} - y_{\pi(2)}) - s_{\pi(1)}(x_{\pi(1)} - y_{\pi(1)}), \\
&\cdots \\
\beta_{n-q+p-2} &= s_{\pi(n-q+p-1)}(x_{\pi(n-q+p-1)} - y_{\pi(n-q+p-1)}) \\
&\quad - s_{\pi(n-q+p-2)}(x_{\pi(n-q+p-2)} - y_{\pi(n-q+p-2)}), \\
\beta_{n-q+p-1} &= -s_{\pi(n-q+p-1)}(x_{\pi(n-q+p-1)} - y_{\pi(n-q+p-1)}) \\
&\quad + (\sum_{k=n-q+p}^{n} s_{\pi(k)}(x_{\pi(k)} - y_{\pi(k)}) - 1)/(q - p), \\
\beta_j &= s_{\pi(j)}(x_{\pi(j)} - y_{\pi(j)}) \\
&\quad + (1 - \sum_{k=n-q+p}^{n} s_{\pi(k)}(x_{\pi(k)} - y_{\pi(k)}))/(q - p),
\end{aligned}
$$

for $j = n - q + p, \cdots, n$. Next, we show that such an integer p exists. In fact, for $p = q - 1$, if $\beta_{n-2} \geq 0$, then it is obvious that $\beta_k \geq 0$ for all k. Otherwise, since $\mu > 1$, there exists $1 \leq p_0 \leq q - 2$ such that

$$
\begin{aligned}
0 \leq &-s_{\pi(n-q+p_0-1)}(x_{\pi(n-q+p_0-1)} - y_{\pi(n-q+p_0-1)}) \\
&+ (\sum_{k=n-q+p_0}^{n} s_{\pi(k)}(x_{\pi(k)} - y_{\pi(k)}) - 1)/(q - p_0)
\end{aligned}
$$

and

$$
\begin{aligned}
0 > &-s_{\pi(n-q+p_0)}(x_{\pi(n-q+p_0)} - y_{\pi(n-q+p_0)}) \\
&+ (\sum_{k=n-q+p_0+1}^{n} s_{\pi(k)}(x_{\pi(k)} - y_{\pi(k)}) - 1)/(q - p_0 - 1),
\end{aligned}
$$

and hence,

$$
\begin{aligned}
&s_{\pi(n-q+p_0)}(x_{\pi(n-q+p_0)} - y_{\pi(n-q+p_0)}) \\
&+ (1 - \sum_{k=n-q+p_0}^{n} s_{\pi(k)}(x_{\pi(k)} - y_{\pi(k)}))/(q - p_0) \geq 0.
\end{aligned}
$$

Then p is equal to p_0. It is clear that $\beta_k \geq 0$ for all k and that

$$\sum_{k=0}^{n} \beta_k = 1 \text{ and } x = \sum_{k=0}^{n} \beta_k y^k,$$

where y^k is as defined in **Definition 5.1.1** for $k = 0, 1, \cdots, n$. Thus

$$x \in T_1(y, \pi, s, p).$$

From the above conclusions, the lemma follows immediately.

END

Lemma 5.1.3. For σ^1 and σ^2 in T_1, the intersection of σ^1 and σ^2 is either a common face of both σ^1 and σ^2 or empty.

Proof. The proof is similar to the one of **Lemma 4.1.4**.

END

Theorem 5.1.4. T_1 is a triangulation of $C^n(m)$.

Proof. From **Definition 5.1.1**, **Lemma 5.1.2**, and **Lemma 5.1.3**, the theorem follows immediately.

END

We call this simplicial subdivision the T_1-triangulation of $C^n(m)$. It is illustrated in Figure 5.1 for $n = 3$ and $m = 2$.

Let N_0 denote the index set $\{0, 1, \cdots, n\}$. Take arbitrarily an interior point $x^0 \in S^n$. For a proper subset H of N_0, let us define the set $S^n(H)$ by

$$S^n(H) = \{x \in S^n \mid x_i = 0 \text{ for all } i \in H\}.$$

Then the orthogonal projection vector of x^0 on $S^n(H)$, $v(H)$, is defined by

$$v_i(H) = \begin{cases} 0 & \text{if } i \in H, \\ x_i^0 + (1 - \sum_{j \notin H} x_j^0)/|N_0 \backslash H| & \text{if } i \notin H, \end{cases}$$

for $i = 0, 1, \cdots, n$. Next, let us define the $(n + 1)$-vector h^i by

$$h^i = v(\{i\}) - x^0$$

for $i = 0, 1, \cdots, n$. For a proper subset H of N_0, let the set $A(H)$ be given by

$$A(H) = \left\{ x \in S^n \;\middle|\; \begin{array}{l} x = x^0 + \sum_{j \in H} \lambda_j h^j, \\ \lambda_j \geq 0 \text{ for all } j \in H \end{array} \right\}.$$

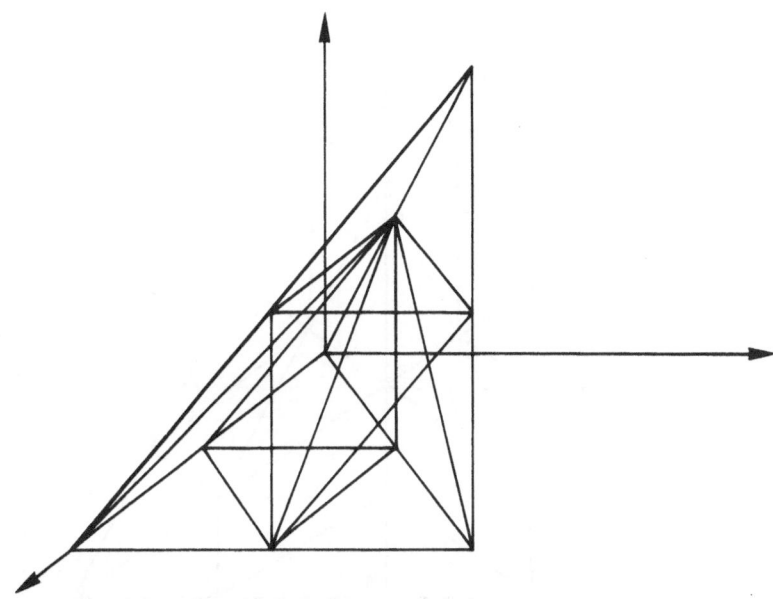

Figure 5.1: The T_1-triangulation of $C^n(m)$ for $n = 3$ and $m = 2$.

It is obvious that the union of the sets $A(H)$ over all subsets H with $|H| = n$ is equal to S^n. Take a permutation $\gamma(H) = (\gamma(1), \gamma(2), \cdots, \gamma(n))$ of elements of H. Then let us define the set $A(\gamma(H))$ by

$$A(\gamma(H)) = \left\{ x \in S^n \;\middle|\; \begin{array}{l} x = x^0 + \sum_{j=1}^n \lambda_j w^j, \\ 1 \geq \lambda_1 \geq \lambda_2 \geq \cdots \geq \lambda_n \geq 0 \end{array} \right\},$$

where

$$w^1 = v(\{\gamma(1)\}) - x^0$$

and

$$w^j = v(\{\gamma(1), \cdots, \gamma(j)\}) - v(\{\gamma(1), \cdots, \gamma(j-1)\})$$

for $j = 2, \cdots, n$. It is obvious that the set $A(H)$ is equal to the union of the sets $A(\gamma(H))$ over all permutations of the elements in H. Further, it is easily seen that the set $A(\gamma(H))$ is homeomorphic to $C^n(m)$. Let W denote the $(n + 1) \times n$ matrix with the i-th column equal to w^i for $i = 1, 2, \cdots, n$. Then

$$A(\gamma(H)) = \left\{ x^0 + m^{-1} W x \mid x \in C^n(m) \right\}.$$

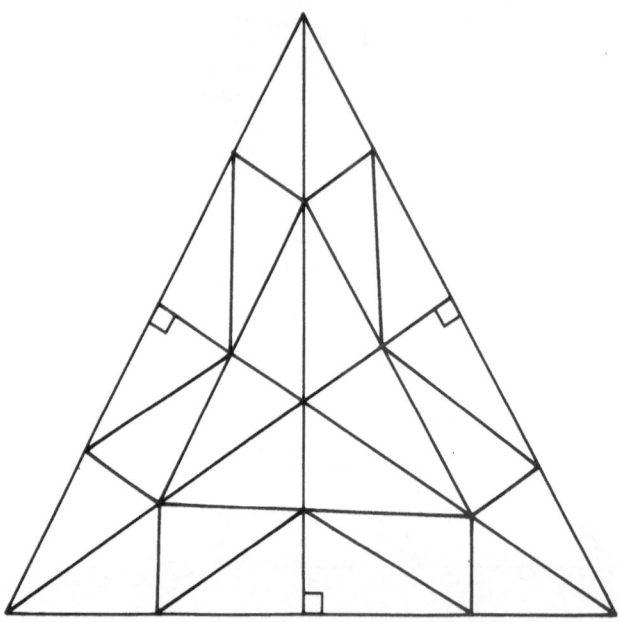

Figure 5.2: The T_1-triangulation of S^n for $n = 2$ and $m = 2$ with $x^0 = (1/3, 1/3, 1/3)^{\mathsf{T}}$.

Thus, the T_1-triangulation of $C^n(m)$ induces a simplicial subdivision of $A(\gamma(H))$ after the same transformation and translation. It is denoted by $T_1(\gamma(H))$. For $H \subset N_0$ with $|H| = n$, a simplicial subdivision of the set $A(H)$, $T_1(H)$, can be derived from the union of $T_1(\gamma(H))$ over all permutations of elements of H. Finally, a triangulation of S^n is obtained from the union of $T_1(H)$ over all $H \subset N_0$ with $|H| = n$. We call this simplicial subdivision the T_1-triangulation of the unit simplex with grid size m^{-1}. It is illustrated in Figure 5.2 for $n = 2$ and $m = 2$. It is clear that the T_1-triangulation induces similarly to the D_1-triangulation a simplicial subdivision of each of the subsets $A(H)$ for H being a proper subset of N_0. These sets underlie the $(n + 1)$-ray variable dimension method. This algorithm was introduced by van der Laan and Talman in [108] using Sperner's labelling rule for computing fixed points or equilibria. We remark that this algorithm con't be used to compute stationary points of a function on S^n. It is refered to Doup [24] for various integer and vector labelling rules. In the next chapter we introduce a version of the D_1-triangulation of $C^n(m)$ which is suitable for every known variable dimension algorithm on S^n.

5.2 Pivot Rules of the T_1-Triangulation

In this section, we give the pivot rules of the T_1-triangulation of $C^n(m)$. Let

$$\sigma = T_1(y, \pi, s, p)$$

be a simplex of the T_1-triangulation of $C^n(m)$ with vertices y^0, y^1, $\cdots$, y^n. We want to obtain the parameters of a simplex of the T_1-triangulation, say

$$\bar{\sigma} = T_1(\bar{y}, \bar{\pi}, \bar{s}, \bar{p}),$$

such that all vertices of σ are also vertices of $\bar{\sigma}$ except vertex y^i, unless the facet of σ opposite y^i lies in the boundary of $C^n(m)$. **Table 5.2.1** describes how $\bar{y}$, $\bar{\pi}$, $\bar{s}$, and $\bar{p}$ are determined from y, π, s, p, and i. From this table, it is easy to obtain the vertices of $\bar{\sigma}$, in particular the vertex opposite to its facet shared with σ.

In this table, q^* is the nonnegative integer such that $\bar{\pi}(k) \in K(\bar{y}, \bar{s})$ for $k = n - q^* + 1, \cdots, n$ and $\bar{\pi}(n - q^*) \notin K(\bar{y}, \bar{s})$. Moreover, if $\bar{y}_{\pi(n-1)} = \bar{y}_{\pi(n)}$ and $\bar{s}_{\pi(n-1)} = \bar{s}_{\pi(n)}$, $p^* = 0$; otherwise $p^* = p$. Further,

$$\hat{q} = \begin{cases} q & \text{if } p^* = p, \\ 1 & \text{if } p^* = 0, \end{cases}$$

and

$$i^* = \begin{cases} n & \text{if } i = n - 1, \\ n - 1 & \text{if } i = n. \end{cases}$$

Table 5.2.1(1): The Pivot Rules of the T_1-Triangulation

i	q	p, n	y, π, s, i	$\bar{y}$	$\bar{s}$	$\bar{\pi}$	$\bar{p}$	$\bar{q}$
0	$q=1$		$y_{\pi(1)}=0$ or $y_{\pi(1)}=m$	case(1)	$s-2s_{\pi(1)}u^{\pi(1)}$	π		
		$n=2$	otherwise	y			p	$q+1$
		$n>2$						q
$1 \le i < n-2$			$y_{\pi(i)}=y_{\pi(i+1)}$, $s_{\pi(i)}=s_{\pi(i+1)}$	BD(1)				
			otherwise	y	s	$(\pi(1),\ldots,\pi(i+1),\pi(i),\ldots,\pi(n))$	p	q
$1 \le i = n-2$			$y_{\pi(i)}=y_{\pi(i+1)}$, $s_{\pi(i)}=s_{\pi(i+1)}$	BD(2)				
			otherwise	y	s	$(\pi(1),\ldots,\pi(i+1),\pi(i),\ldots,\pi(n))$	p	$q+1$
$1 \le i = n-1$				BD(3)				
n			$y_{\pi(n)}+s_{\pi(n)}=m$	case(2)				
			otherwise	$y+2s_{\pi(n)}u^{\pi(n)}$	$s-2s_{\pi(n)}u^{\pi(n)}$	π	q^*-1	q^*
0	$2 \le q < n-1$		$y_{\pi(1)}=0$ or $y_{\pi(1)}=m$	case(3)				
			otherwise	y	$s-2s_{\pi(1)}u^{\pi(1)}$	π	p	q
	$2 \le q = n-1$	$y_{\pi(1)}=0$ or $y_{\pi(1)}=m$	$\pi(1) \in J(y)$	case(4)				
		otherwise						
		$p=0$	$\pi(1) \in I^+(y) \cup I^-(y)$	y	$s-2s_{\pi(1)}u^{\pi(1)}$	π	p	$q+1$
		$p \ge 1$	$\pi(1) \in I^+(y) \cup I^-(y)$	y	$s-2s_{\pi(1)}u^{\pi(1)}$	π	$p+1$	$q+1$
$1 \le i \le n-q-2$	$2 \le q < n$		$y_{\pi(i)}=y_{\pi(i+1)}$, $s_{\pi(i)}=s_{\pi(i+1)}$	BD(4)				
			otherwise	y	s	$(\pi(1),\ldots,\pi(i+1),\pi(i),\ldots,\pi(n))$	p	q

Table 5.2.1(2): The Pivot Rules of the T_1-Triangulation

i	q	p, n	y, π, s, i	$\bar y$	$\bar s$	$\bar\pi$	$\bar p$	$\bar q$
$1 \leq i = n-q-1$	$2 \leq q < n$	$p = 0$	$\pi(i) \in K(y,s)$	y	s	$(\pi(1),\ldots,\pi(i+1),\pi(i),\ldots,\pi(n))$	p	$q+1$
			$y_{\pi(i)} = y_{\pi(i+1)}$, $s_{\pi(i)} = s_{\pi(i+1)}$, otherwise	BD(5)				
		$p \geq 1$	$\pi(i) \in K(y,s)$	y	s	$(\pi(1),\ldots,\pi(i+1),\pi(i),\ldots,\pi(n))$	p	q
				y	s	$(\pi(1),\ldots,\pi(i+1),\pi(i),\ldots,\pi(n))$	$p+1$	$q+1$
			$y_{\pi(i)} = y_{\pi(i+1)}$, $s_{\pi(i)} = s_{\pi(i+1)}$, otherwise	BD(6)				
$n-q$		$p = 0$		y	s	$(\pi(1),\ldots,\pi(i+1),\pi(i),\ldots,\pi(n))$	p	q
		$p = 1$		y	s	$\bar\pi$	$p+1$	q
		$p \geq 2$	$y_{\pi(i)} = y_{\pi(i+1)}$, $s_{\pi(i)} = s_{\pi(i+1)}$	y	s	$\bar\pi$	$p-1$	q
			otherwise	BD(7)				
$n-q < i$		$p = 0$	$y_{\pi(i)} = y_{\pi(n-q)}$, $s_{\pi(i)} = s_{\pi(n-q)}$	y	s	$(\pi(1),\ldots,\pi(i+1),\pi(i),\ldots,\pi(n))$	$p-1$	$q-1$
			otherwise	BD(8)				
$n-q < i < n-q+p-1$				y	s	$(\pi(1),\ldots,\pi(n-q-1),\pi(i),\pi(n-q),\ldots,\pi(i-1),\pi(i+1),\ldots,\pi(n))$	p	$q-1$
$n-q+p-1$				y	s	$(\pi(1),\ldots,\pi(i-1),\pi(i+1),\ldots,\pi(n))$	p	q
		$p \geq 2$		y	s	$\bar\pi$	$p-1$	q

Table 5.2.1(3): The Pivot Rules of the T_1-Triangulation

i	q	p,n	y,π,s,i	$\bar{y}$	$\bar{s}$	$\bar{\pi}$	$\bar{p}$	$\bar{q}$
$n-q+p-1 < i$	$2 \le q < n$	$1 \le p < q-1$		y	s	$(\pi(1),\ldots,\pi(n-q+p-1),\ \pi(i),\ \pi(n-q+p),\ldots,\ \pi(i-1),\pi(i+1),\ldots,\pi(n))$	$p+1$	q
		$1 \le p = q-1$	$y_{\pi(i^*)} + s_{\pi(i^*)} = m$	case(5)				
			$i = n-1$	$y + 2s_{\pi(n)}u^{\pi(n)}$	$s - 2s_{\pi(n)}u^{\pi(n)}$	π	p^*	$\hat{q}$
			$i = n$	$y + 2s_{\pi(n-1)}u^{\pi(n-1)}$	$s - 2s_{\pi(n-1)}u^{\pi(n-1)}$	$(\pi(1),\ldots,\pi(n),\pi(n-1))$	p^*	$\bar{q}$
0	$q = n$	$p = 0$		y	s	π	$p+1$	q
		$p = 1$		y	s	π	$p-1$	q
		$2 \le p$	$y_{\pi(1)} = 0$ or $y_{\pi(1)} = m$	case(6)				$q-1$
			$\pi(1) \in I^+(y) \cup I^-(y)$	y	$s - 2s_{\pi(1)}u^{\pi(1)}$	π	$p-1$	q
			$\pi(1) \in I(y)$	y	$s - 2s_{\pi(1)}u^{\pi(1)}$	π	p	q
$1 \le i$		$p = 0$	$\pi(i) \in I(y)$	y	$s - 2s_{\pi(i)}u^{\pi(i)}$	π	p	$q-1$
			$y_{\pi(i)} = 0$ or $y_{\pi(i)} = m$	case(7)				q
			$\pi(i) \in I^+(y) \cup I^-(y)$	y	$s - 2s_{\pi(i)}u^{\pi(i)}$	$(\pi(i),\pi(1),\ldots,\pi(i-1),\ \pi(i+1),\ldots,\pi(n))$	p	$q-1$
		$i < p-1$		y	s	$(\pi(1),\ldots,\pi(i+1),\ \pi(i),\ldots,\pi(n))$	p	q
		$i = p-1$		y	s	π	$p-1$	$q-1$
		$i > p-1$, $1 \le p < n-1$		y	s	$(\pi(1),\ldots,\pi(p-1),\pi(i),\ \pi(p),\ldots,\pi(i-1),\ \pi(i+1),\ldots,\pi(n))$	$p+1$	q
		$p = n-1$, $i > p-1$	$y_{\pi(i^*)} + s_{\pi(i^*)} = m$	case(8)				
			$i = n-1$	$y + 2s_{\pi(n)}u^{\pi(n)}$	$s - 2s_{\pi(n)}u^{\pi(n)}$	π	p^*	$\hat{q}$
			$i = n$	$y + 2s_{\pi(n-1)}u^{\pi(n-1)}$	$s - 2s_{\pi(n-1)}u^{\pi(n-1)}$	$(\pi(1),\ldots,\pi(n),\pi(n-1))$	p^*	$\bar{q}$

5.3 Comparison of the Triangulations of the Unit Simplex

In the previous chapter it was demonstrated that the D_1-triangulation of R^n is superior to the other well-known triangulations of R^n according to measures of efficiency of triangulations, for example, the number of simplices in a unit cube, the diameter, and the average directional density. From the definition of the T_1-triangulation of $C^n(m)$, it can be seen that the T_1-triangulation is obtained by combining the D_1-triangulation and the J_1-triangulation, i.e., we triangulate all cubes with one of its vertices belonging to D according to the D_1-triangulation. Let $\bar{K}_1$ denote the simplicial subdivision of $C^n(m)$ derived from the restriction of the K_1-triangulation to $C^n(m)$. Next, let $\mathcal{N}(T_1)$ denote the number of simplices of the T_1-triangulation in $C^n(m)$ and $\mathcal{N}(\bar{K}_1)$ the number of simplices of the $\bar{K}_1$-triangulation in $C^n(m)$. It can easily be shown that

$$\lim_{n \to \infty} \lim_{m \to \infty} \mathcal{N}(T_1)/\mathcal{N}(\bar{K}_1) = e - 2.$$

Hence, the T_1-triangulation of the unit simplex is superior to the triangulations that are obtained from the K_1-triangulation or the J_1-triangulation in the same approach as that in the last part of the first section.

Chapter 6

The D_1-Triangulation in Variable Dimension Algorithms on the Unit Simplex

Simplicial variable dimension algorithms to compute fixed points on the unit simplex were initiated by Kuhn in [98]. Kuhn's variable dimension algorithm subdivides the unit simplex according to the Q-triangulation and starts at one of vertices of the unit simplex. It generates a sequence of adjacent simplices with varying dimension and terminates as soon as an approximate solution is yielded. If the accuracy is not good enough, one can restart Kuhn's variable dimension algorithm at one of the vertices of the unit simplex with a finer simplicial subdivision. However, it is obvious that Kuhn's algorithm loses all information about the location of a solution obtained in the previous implementation when restarting. To overcome this drawback, van der Laan and Talman originated a new generation of simplicial variable dimension algorithms on the unit simplex in [108]. Van der Laan and Talman's variable dimension algorithm also subdivides the unit simplex according to the Q-triangulation, but it can start at an arbitrary grid point. Van der Laan and Talman's algorithm leaves this grid point along one out of $n + 1$ different rays. It also generates a sequence of adjacent simplices with varying dimension and terminates in case that an approximate solution is obtained. If the accuracy is not high enough, one can restart the algorithm at the grid point closest to the approximate solution obtained in the previous implementation. In this way the algorithm becomes efficient. However, the $(n + 1)$-ray variable dimension algorithm of van der Laan and Talman in [108] can only be used to compute fixed points or zero points on the unit simplex. To compute stationary points, van der Laan, Talman and Van der Heyden in [122] generalized that algorithm. Also the Q-triangulation underlies this algorithm. In [27] Doup and Talman proposed the $(n + 1)$-ray variable dimension algorithm to compute stationary points on S^n, based on the more efficient V-triangulation. Also other variable dimension algorithms based on the V-triangulation have been proposed. The $(2^{n+1} - 2)$-ray variable dimension method was introduced by Doup, van der Laan and Talman in [26] and the 2-ray variable dimension method by Doup and Talman in [28]. For a survey of all these variable dimension methods on the unit simplex, we refer to Doup [24]. The simplicial subdivisions of the unit simplex that underlie all these algorithms are the Q-triangulation or the V-triangulation. Both these triangulations are based on the K_1-triangulation of R^n. As was seen in Chapter 4, the D_1-triangulation of R^n is superior to the K_1-triangulation according to measures of efficiency such as the number of simplices, the diameter, and the average directional density. Therefore, it is significant to be able to apply the D_1-triangulation to variable dimension algorithms on the unit simplex. Since it is not straightforward to incorporate the D_1-triangulation in variable dimension algorithms on the unit simplex, in this chapter we consider how the D_1-triangulation can be adapted for use in the variable dimension methods mentioned above on the unit simplex. A version of the D_1-triangulation is proposed such that it induces similarly to the D_1-triangulation a simplicial subdivision of each of the subsets, into which a variable dimension algorithm subdivides the unit simplex. One can incorporate this version directly in all known variable dimension methods on the unit simplex. It is expected that the cost of computation can be reduced when using versions of the D_1-triangulation instead of the K_1-triangulation. This chapter is organized as follows. Section 1 introduces the new version of the D_1-triangulation, called the D_{v1}-triangulation. Section 2 gives the pivot rules of the D_{v1}-triangulation. Section 3 and 4 describe how to apply this triangulation to the $(n + 1)$-ray and the $(2^{n+1} - 2)$-ray variable dimension methods on the unit simplex, respectively. This chapter is based on Dang and Talman's [19].

6.1 The D_{v1}-Triangulation

The D_{v1}-triangulation is based on the D_1-triangulation of R^n introduced in Chapter 4. The D_{v1}-triangulation will be used as the underlying simplicial subdivision for variable dimension algorithms on the unit simplex.

Let a positive integer m be given and let N denote the index set $\{1, 2, \cdots, n\}$. Furthermore, let us define the set $C^n(m)$ by

$$C^n(m) = \{x \in R^n \mid m \geq x_1 \geq x_2 \geq \cdots \geq x_n \geq 0\}.$$

The D_{v1}-triangulation subdivides simplicially the set $C^n(m)$. The simplices of this triangulation can be represented by some specific vector y, permutation π, sign vector s, and integer p. Let D denote the set

$$\{y \in C^n(m) \mid \text{all components of } y \text{ are even}\}.$$

Take $y \in D$. Let $1 \leq h \leq n$ be the integer such that for $j = 1, 2, \cdots, h$, it holds that $y_{m_j+1} = y_{m_j+k}$ for $k = 2, 3, \cdots, n_j$, and for $j = 1, 2, \cdots, h-1$, it holds that $y_{m_j+n_j} > y_{m_{j+1}+1}$, where $m_1 = 0$, $n_h = n - m_h$, and $m_j + n_j = m_{j+1}$ for $j = 1, 2, \cdots, h - 1$.

Take a sign vector $s = (s_1, s_2, \cdots, s_n)^{\mathsf{T}}$ such that

$$s_{m_j+1} = \cdots = s_{m_j+k_j} = 1$$

and

$$s_{m_j+k_j+1} = \cdots = s_{m_j+n_j} = -1$$

for $j = 1, 2, \cdots, h$, where $k_1 = 0$ if $y_1 = m$, $k_h = n_h$ if $y_n = 0$, and $0 \leq k_j \leq n_j$ for $j = 1, 2, \cdots, h$.

For $1 \leq j \leq h$ and $1 \leq k \leq k_j$, let the n-vector $g(m_j + k)$ be given by

$$g_i(m_j + k) = \begin{cases} 1 & \text{if } m_j + 1 \leq i \leq m_j + k, \\ 0 & \text{otherwise,} \end{cases}$$

for $i = 1, 2, \cdots, n$.

For $1 \leq j \leq h$ and $k_j + 1 \leq k \leq n_j$, let the n-vector $g(m_j + k)$ be given by

$$g_i(m_j + k) = \begin{cases} -1 & \text{if } m_j + k \leq i \leq m_j + n_j, \\ 0 & \text{otherwise,} \end{cases}$$

for $i = 1, 2, \cdots, n$.

Take an integer p such that $0 \leq p \leq n - 1$ and if $h = 1$ and $k_1 = 0$ or n then $p = 0$.

Finally, take a permutation π of the elements of N such that for $1 \leq j \leq h$,

$$\pi^{-1}(m_j + 1) > \pi^{-1}(m_j + 2) > \cdots > \pi^{-1}(m_j + k_j)$$

and

$$\pi^{-1}(m_j + k_j + 1) < \pi^{-1}(m_j + k_j + 2) < \cdots < \pi^{-1}(m_j + n_j)$$

and when $p \geq 1$, there exists no $1 \leq j \leq h$ such that $m_j + 1 \leq \pi(k) \leq m_j + k_j$ for all $p \leq k \leq n$ or $m_j + k_j + 1 \leq \pi(k) \leq m_j + n_j$ for all $p \leq k \leq n$.

For $i = 1, 2, \cdots, n$, let u^i denote the i-th unit vector in R^n.

Definition 6.1.1. For the vector y, the permutation π, the sign vector s, and the number p given as above, the vectors $y^0, y^1, \cdots, y^n$ are given as follows.

If $p = 0$, then $y^0 = y$ and

$$y^k = y + g(\pi(k)), \quad k = 1, 2, \cdots, n.$$

If $p \geq 1$, then $y^0 = y + s$,

$$y^k = y^{k-1} - s_{\pi(k)} u^{\pi(k)}, \quad k = 1, 2, \cdots, p - 1, \text{ and}$$
$$y^k = y + g(\pi(k)), \quad k = p, p + 1, \cdots, n.$$

Let $y^0, y^1, \cdots, y^n$ be obtained from the above definition. Then it is obvious that they are affinely independent. Thus their convex hull is a simplex with vertices y^0, y^1, $\cdots$, y^n. Let us denote this simplex by $D_{v1}(y, \pi, s, p)$. Let D_{v1} be the set of simplices $D_{v1}(y, \pi, s, p)$ for all y, π, s, and p given as above. We will show that D_{v1} is a simplicial subdivision of $C^n(m)$.

For the given y and s as above, let α be defined by $\alpha = \sum_{j=1}^{h} \alpha_j$, where

$$\alpha_j = \begin{cases} 2 & \text{if } 0 < k_j < n_j, \\ 1 & \text{otherwise,} \end{cases}$$

for $j = 1, 2, \cdots, h$.

For the given y, π, s, and p as above, let

$$r_j = |\, \{m_j + k \mid 1 \le k \le k_j\} \cap \{\pi(k) \mid 1 \le k \le p - 1\} \,|$$

and

$$q_j = |\, \{m_j + k \mid k_j + 1 \le k \le n_j\} \cap \{\pi(k) \mid 1 \le k \le p - 1\} \,|$$

for $j = 1, 2, \cdots, h$.

Lemma 6.1.2. The union of all $\sigma \in D_{v1}$ is equal to $C^n(m)$.

Proof. Clearly, every $\sigma \in D_{v1}$ is a subset of $C^n(m)$. Let $x \in C^n(m)$ be arbitrary. Then $x \in D_{v1}(y, \pi, s, p)$ with y, π, s, and p determined as follows.

Take the vector y equal to

$$y_i = \begin{cases} \lfloor x_i \rfloor + 1 & \text{if } \lfloor x_i \rfloor \text{ is odd and } \lfloor x_i \rfloor < m, \\ \lfloor x_i \rfloor - 1 & \text{if } \lfloor x_i \rfloor \text{ is odd and } \lfloor x_i \rfloor = m, \\ \lfloor x_i \rfloor & \text{otherwise,} \end{cases}$$

for $i = 1, 2, \cdots, n$ and the sign vector s equal to

$$s_i = \begin{cases} -1 & \text{if } \lfloor x_i \rfloor \text{ is odd and } \lfloor x_i \rfloor < m \\ & \quad \text{or } \lfloor x_i \rfloor \text{ is even and } \lfloor x_i \rfloor = m, \\ 1 & \text{otherwise,} \end{cases}$$

for $i = 1, 2, \cdots, n$. It is obvious that $y \in D$. Take the permutation π such that

$$s_{\pi(1)}\big(x_{\pi(1)} - y_{\pi(1)}\big) \le s_{\pi(2)}\big(x_{\pi(2)} - y_{\pi(2)}\big) \le \cdots \le s_{\pi(n)}\big(x_{\pi(n)} - y_{\pi(n)}\big)$$

and for $1 \le j \le h$,

$$\pi^{-1}(m_j + 1) > \pi^{-1}(m_j + 2) > \cdots > \pi^{-1}(m_j + k_j)$$

and

$$\pi^{-1}(m_j + k_j + 1) < \pi^{-1}(m_j + k_j + 2) < \cdots < \pi^{-1}(m_j + n_j).$$

Suppose $\alpha = 1$. Then take $p = 0$. Now, let $y^0 = y$ and

$$y^k = y + g(\pi(k)), \quad k = 1, 2, \cdots, n,$$

and if $s_1 = 1$, let

$$
\begin{aligned}
\beta_0 &= 1 - s_1(x_1 - y_1), \\
\beta_1 &= s_n(x_n - y_n), \\
\beta_2 &= s_{n-1}(x_{n-1} - y_{n-1}) - s_n(x_n - y_n), \\
&\cdots, \\
\beta_n &= s_1(x_1 - y_1) - s_2(x_2 - y_2),
\end{aligned}
$$

and if $s_1 = -1$, let

$$
\begin{aligned}
\beta_0 &= 1 - s_n(x_n - y_n), \\
\beta_1 &= s_1(x_1 - y_1), \\
\beta_2 &= s_2(x_2 - y_2) - s_1(x_1 - y_1), \\
&\cdots, \\
\beta_n &= s_n(x_n - y_n) - s_{n-1}(x_{n-1} - y_{n-1}).
\end{aligned}
$$

Obviously, $\sum_{k=0}^{n} \beta_k = 1$, $\beta_k \ge 0$ for all k, and

$$x = \sum_{k=0}^{n} \beta_k y^k.$$

Hence,

$$x \in D_{v1}(y, \pi, s, p).$$

When $\alpha = n$, the proof is the same as that of **Lemma 4.1.3**. Now suppose that $1 < \alpha < n$. For $1 \le j \le h$, let

$$
\mu_j = \begin{cases}
s_{m_j+1}(x_{m_j+1} - y_{m_j+1}) + s_{m_j+n_j}(x_{m_j+n_j} - y_{m_j+n_j}) \\
\qquad \text{if } \alpha_j = 2, \\
s_{m_j+1}(x_{m_j+1} - y_{m_j+1}) \\
\qquad \text{if } \alpha_j = 1 \text{ and } s_{m_j+1} = 1, \\
s_{m_j+n_j}(x_{m_j+n_j} - y_{m_j+n_j}) \\
\qquad \text{if } \alpha_j = 1 \text{ and } s_{m_j+1} = -1.
\end{cases}
$$

Let $\mu = \sum_{j=1}^{h} \mu_j$.

Suppose that $\mu \leq 1$. Take $p = 0$ and let $y^0 = y$ and

$$y^k = y + g(\pi(k)), \quad k = 1, 2, \cdots, n.$$

For $1 \leq k \leq n$, let

$$\beta_k = \begin{cases} s_{\pi(k)}\left(x_{\pi(k)} - y_{\pi(k)}\right) \\ \quad \text{if } \pi(k) = m_j + k_j \text{ and } k_j \geq 1 \text{ or } \pi(k) = \\ \quad m_j + k_j + 1 \text{ and } k_j \leq n_j - 1 \text{ for some } 1 \leq j \leq h, \\ s_{\pi(k)}\left(x_{\pi(k)} - y_{\pi(k)}\right) - s_{\pi(k)+1}\left(x_{\pi(k)+1} - y_{\pi(k)+1}\right) \\ \quad \text{if } 1 \leq \pi(k) - m_j < k_j \text{ for some } 1 \leq j \leq h, \\ s_{\pi(k)}\left(x_{\pi(k)} - y_{\pi(k)}\right) - s_{\pi(k)-1}\left(x_{\pi(k)-1} - y_{\pi(k)-1}\right) \\ \quad \text{if } k_j + 1 < \pi(k) - m_j \leq n_j \text{ for some } 1 \leq j \leq h, \end{cases}$$

and let $\beta_0 = 1 - \mu$. Then it is clear that $\beta_k \geq 0$ for all k, $\sum_{k=0}^{n} \beta_k = 1$, and

$$x = \sum_{k=0}^{n} \beta_k y^k.$$

Thus

$$x \in D_{v1}(y, \pi, s, p).$$

Now suppose $\mu > 1$. Let p_{max} denote the largest $1 \leq p \leq n - 1$ such that there exists no $1 \leq j \leq h$ such that $m_j + 1 \leq \pi(k) \leq m_j + k_j$ for all $p \leq k \leq n$ or $m_j + k_j + 1 \leq \pi(k) \leq m_j + n_j$ for all $p \leq k \leq n$. Next we show that we can take the integer p, $1 \leq p \leq p_{max}$, such that the following system has a nonnegative solution,

$$\begin{aligned} \beta_0 &= s_{\pi(1)}\left(x_{\pi(1)} - y_{\pi(1)}\right), \\ \beta_1 &= s_{\pi(2)}\left(x_{\pi(2)} - y_{\pi(2)}\right) - s_{\pi(1)}\left(x_{\pi(1)} - y_{\pi(1)}\right), \\ &\quad \cdots, \\ \beta_{p-2} &= s_{\pi(p-1)}\left(x_{\pi(p-1)} - y_{\pi(p-1)}\right) \\ &\quad - s_{\pi(p-2)}\left(x_{\pi(p-2)} - y_{\pi(p-2)}\right), \\ \beta_{p-1} &= -s_{\pi(p-1)}\left(x_{\pi(p-1)} - y_{\pi(p-1)}\right) \\ &\quad + \left(\sum_{i=p}^{n} \lambda_{\pi(i)} - 1\right)/(c(p) - 1), \end{aligned}$$

$$\beta_k = \begin{cases} s_{\pi(k)}\big(x_{\pi(k)} - y_{\pi(k)}\big) - s_{\pi(k)+1}\big(x_{\pi(k)+1} - y_{\pi(k)+1}\big) \\ \qquad \text{if } 1 \leq \pi(k) - m_j < k_j - r_j \text{ for some } 1 \leq j \leq h, \\ s_{\pi(k)}\big(x_{\pi(k)} - y_{\pi(k)}\big) - \big(\sum_{i=p}^{n} \lambda_{\pi(i)} - 1\big)/(c(p) - 1) \\ \qquad \text{if } 1 \leq \pi(k) - m_j = k_j - r_j \\ \qquad \text{or } k_j + q_j + 1 = \pi(k) - m_j \leq n_j \text{ for some } 1 \leq j \leq h, \\ s_{\pi(k)}\big(x_{\pi(k)} - y_{\pi(k)}\big) - s_{\pi(k)-1}\big(x_{\pi(k)-1} - y_{\pi(k)-1}\big) \\ \qquad \text{if } k_j + q_j + 1 < \pi(k) - m_j \leq n_j \text{ for some } 1 \leq j \leq h, \end{cases}$$

for $k = p, p+1, \cdots, n$, where for $i = p, \cdots, n$,

$$\lambda_{\pi(i)} = \begin{cases} 0 & \text{if } 1 < \pi(i) - m_j \leq k_j - r_j \\ & \text{or } k_j + q_j + 1 \leq \pi(i) - m_j < n_j \\ & \text{for some } 1 \leq j \leq h, \\ s_{m_j+1}(x_{m_j+1} - y_{m_j+1}) & \text{if } 1 = \pi(i) - m_j \leq k_j - r_j \\ & \text{for some } 1 \leq j \leq h, \\ s_{m_j+n_j}(x_{m_j+n_j} - y_{m_j+n_j}) & \text{if } k_j + q_j + 1 \leq \pi(i) - m_j = n_j \\ & \text{for some } 1 \leq j \leq h, \end{cases}$$

and

$$c(p) = \sum_{j=1}^{h} c_j(p)$$

with

$$c_j(p) = \begin{cases} 0 & \text{if } r_j = k_j \text{ and } q_j = n_j - k_j, \\ 2 & \text{if } r_j < k_j \text{ and } q_j < n_j - k_j, \\ 1 & \text{otherwise}, \end{cases}$$

for $j = 1, 2, \cdots, h$. If $\beta_{p-1} \geq 0$ in case $p = p_{max}$, it is clear that $\beta_k \geq 0$ for all k, and we take p equal to p_{max}. Otherwise, since $\mu > 1$, there exists $1 \leq p_0 \leq p_{max} - 1$ such that

$$0 \leq \ -s_{\pi(p_0-1)}\big(x_{\pi(p_0-1)} - y_{\pi(p_0-1)}\big) \\ + \big(\sum_{i=p_0}^{n} \lambda_{\pi(i)} - 1\big)/(c(p_0) - 1)$$

and

$$0 > \ -s_{\pi(p_0)}\big(x_{\pi(p_0)} - y_{\pi(p_0)}\big) \\ + \big(\sum_{i=p_0+1}^{n} \lambda_{\pi(i)} - 1\big)/(c(p_0 + 1) - 1),$$

and hence,

$$s_{\pi(p_0)}\big(x_{\pi(p_0)} - y_{\pi(p_0)}\big) - \big(\sum_{i=p_0}^{n} \lambda_{\pi(i)} - 1\big)/(c(p_0) - 1) \geq 0.$$

Thus, if $p = p_0$ then $\beta_k \geq 0$ for all k. We take p equal to p_0. Obviously, $\sum_{k=0}^{n} \beta_k = 1$. Let $y^0 = y + s$, for $k = 1, 2, \cdots, p-1$, let

$$y^k = y^{k-1} - s_{\pi(k)} u^{\pi(k)},$$

and for $k = p, p+1, \cdots, n$, let

$$y^k = y + g(\pi(k)).$$

We easily obtain that $x = \sum_{k=0}^{n} \beta_k y^k$. Therefore,

$$x \in D_{v1}(y, \pi, s, p).$$

From the above conclusions, the lemma follows immediately.

END

Lemma 6.1.3. For σ^1 and σ^2 in D_{v1}, the intersection of σ^1 and σ^2 is either a common face of both σ^1 and σ^2 or empty.

Proof. The proof is similar to the one of **Lemma 4.1.4**.

END

Theorem 6.1.4. D_{v1} is a triangulation of $C^n(m)$.

Proof. From the definition, **Lemma 6.1.2** and **Lemma 6.1.3**, the theorem follows immediately.

END

This simplicial subdivision is called the D_{v1}-triangulation of $C^n(m)$. The D_{v1}-triangulation of $C^n(m)$ is illustrated in Figure 6.1 for $n = 3$ and $m = 2$.

The D_{v1}-triangulation of $C^n(m)$ is such that the set $C^n(m)$ is subdivided into simplices according to the D_1-triangulation of R^n with grid size m^{-1} except along the boundary of $C^n(m)$. Moreover, contrary to the T_1-triangulation introduced in Chapter 5, every t-dimensional face of $C^n(m)$ is also simplicially sudivided according to the D_1-triangulation of R^t, $1 \leq t \leq n-1$. Therefore, the D_{v1}-triangulation of $C^n(m)$ is superior to other triangulations of this set according to the conclusions in Chapter 4.

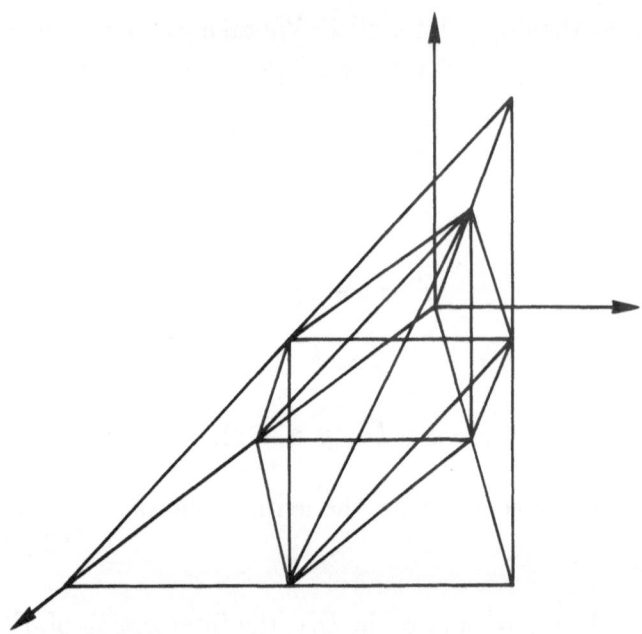

Figure 6.1: The D_{v1}-triangulation of $C^n(m)$ for $n = 3$ and $m = 2$.

6.2   Pivot Rules of the D_{v1}-Triangulation

Let $\sigma = D_{v1}(y, \pi, s, p)$ be a simplex of the D_{v1}-triangulation of $C^n(m)$ with vertices y^0, y^1, $\cdots$, y^n. We want to obtain the parameters of the simplex $\bar{\sigma} = D_{v1}(\bar{y}, \bar{\pi}, \bar{s}, \bar{p})$ sharing with σ the facet opposite to the vertex y^i, $i \in \{0, 1, \cdots, n\}$, of σ, in case that this facet of σ does not lie in the boundary of $C^n(m)$. In **Table 6.2.1**, we show how $\bar{y}$, $\bar{\pi}$, $\bar{s}$ and $\bar{p}$ depend on y, π, s, p and i. From this table, it is easy to obtain each vertex of $\bar{\sigma}$, and in particular the vertex opposite to the facet shared with σ.

In this table, π^* and p^* denote that if, for some $1 \leq j \leq \bar{h}$,

$$1 \leq \pi(k) - \bar{m}_j \leq \bar{k}_j \text{ for } k = n - 1, n$$

or

$$\bar{k}_j + 1 \leq \pi(k) - \bar{m}_j \leq \bar{n}_j \text{ for } k = n - 1, n,$$

then

$$\pi^* = (\pi(1), \cdots, \pi(n - 2), \pi(n), \pi(n - 1))$$

and

$$
p^* = \begin{cases}
k & \text{if there exists } 1 \le k \le n-2 \text{ such that } 1 \le \pi(l) - \bar{m}_j \le \bar{k}_j \\
& \text{for all } k < l \le n \text{ and } \pi(k) \text{ doesn't or} \\
& \bar{k}_j + 1 \le \pi(l) - \bar{m}_j \le \bar{n}_j \\
& \text{for all } k < l \le n \text{ and } \pi(k) \text{ doesn't,} \\
0 & \text{otherwise;}
\end{cases}
$$

otherwise, $\pi^* = \pi$ and $p^* = p$, where $\bar{h}$, $\bar{m}_j$, $\bar{k}_j$, and $\bar{n}_j$ satisfy that for $j = 1, 2, \cdots, \bar{h}$, it holds that $\bar{y}_{\bar{m}_j+1} = \bar{y}_{\bar{m}_j+k}$ for $k = 2, 3, \cdots, \bar{n}_j$, and for $j = 1, 2, \cdots, \bar{h} - 1$, it holds that $\bar{y}_{\bar{m}_j+\bar{n}_j} > \bar{y}_{\bar{m}_{j+1}+1}$, where $\bar{m}_1 = 0$, $\bar{n}_{\bar{h}} = n - \bar{m}_{\bar{h}}$, and $\bar{m}_j + \bar{n}_j = \bar{m}_{j+1}$ for $j = 1, 2, \cdots, \bar{h} - 1$, and for $j = 1, 2, \cdots, \bar{h}$, it holds that $\bar{s}_{\bar{m}_j+1} = \cdots = \bar{s}_{\bar{m}_j+\bar{k}_j} = 1$ and $\bar{s}_{\bar{m}_j+\bar{k}_j+1} = \cdots = \bar{s}_{\bar{m}_j+\bar{n}_j} = -1$, where $\bar{k}_1 = 0$ if $\bar{y}_1 = m$, and $\bar{k}_{\bar{h}} = \bar{n}_{\bar{h}}$ if $\bar{y}_n = 0$.

Table 6.2.1(1). The Pivot Rules of the D_{v1}-Triangulation

i	p		$\bar{y}$	$\bar{s}$	$\bar{\pi}$	$\bar{p}$
0	0	$\alpha=1, s_1=1$ $y_1=m-1$	BD(1)			
		otherwise	$y+2s_1u^1$	$s-2s_1u^1$	π	$n-1$
		$\alpha=1, s_1=-1$	$y+2s_nu^n$	$s-2s_nu^n$	π	$n-1$
		$\alpha\geq 2$	y	s	π	$p+1$
0	1		y	s	π	$p-1$
0	$p\geq 2$	$y_{\pi(1)}=0$ or m	BD(2)			
		otherwise	y	$s-2s_{\pi(1)}u^{\pi(1)}$	π	p
$1\leq i\leq n$	0	$y_{\pi(i)}=0$ or m	BD(3)			
		otherwise	y	$s-2s_{\pi(i)}u^{\pi(i)}$	π	p
$1\leq i<p-1$		$1\leq\pi(i)-m_j=k_j$ or $k_j+1=\pi(i)-m_j$ for some $1\leq j\leq h$	BD(4)			
		for some $1\leq j\leq h$, $1\leq\pi(k)-m_j\leq k_j$ for $k=i,i+1$ or $k_j+1\leq\pi(k)-m_j\leq n_j$ for $k=i,i+1$	BD(5)			
		otherwise	y	s	$(\pi(1),\ldots,\pi(i+1),$ $\pi(i),\ldots,\pi(n))$	p
$i=p-1$	$p\geq 2$		y	s	π	$p-1$

Table 6.2.1(2). The Pivot Rules of the D_{v1}-Triangulation

i	p		$\bar{y}$	$\bar{s}$	$\bar{\pi}$	$\bar{p}$
$p \leq i \leq n$	$1 \leq p < n-1$	for some $1 \leq j \leq h$, $1 \leq \pi(k) - m_j \leq k_j$ for $p \leq k \leq n$ and $k \neq i$ or $k_j + 1 \leq \pi(k) - m_j \leq n_j$ for $p \leq k \leq n$ and $k \neq i$: $i < n$ and either $y_{\pi(n)} \neq m - 1$ or $s_{\pi(n)} \neq 1$	$y + 2s_{\pi(n)}$ $u^{\pi(n)}$	$s - 2s_{\pi(n)}$ $u^{\pi(n)}$	$(\pi(1),\ldots,\pi(p-1),$ $\pi(i),\pi(p),\ldots,\pi(i-1),$ $\pi(i+1),\ldots,\pi(n))$	$n-1$
		$i = n$ and either $y_{\pi(n-1)} \neq m - 1$ or $s_{\pi(n-1)} \neq 1$	$y + 2s_{\pi(n-1)}$ $u^{\pi(n-1)}$	$s - 2s_{\pi(n-1)}$ $u^{\pi(n-1)}$	$(\pi(1),\ldots,\pi(p-1),$ $\pi(n),\pi(p),\ldots,\pi(n-1))$	$n-1$
		otherwise	BD(6)			
	$1 \leq p \equiv n-1$	for some $1 \leq j \leq h$, $1 \leq \pi(i) - m_j = k_j$ $-r_j$ or $k_j + g_j + 1$ $= \pi(i) - m_j \leq n_j$	y	s	$(\pi(1),\ldots,\pi(p-1),$ $\pi(i),\pi(p),\ldots,\pi(i-1),$ $\pi(i+1),\ldots,\pi(n))$	$p+1$
		otherwise	BD(7)			
		$i = n-1$ and either $s_{\pi(n)} \neq 1$ or $y_{\pi(n)} \neq m - 1$	$y + 2s_{\pi(n)}$ $u^{\pi(n)}$	$s - 2s_{\pi(n)}$ $\cdot u^{\pi(n)}$	π	p^*
		$i = n$ and either $y_{\pi(n-1)} \neq m - 1$ or $s_{\pi(n-1)} \neq 1$	$y + 2s_{\pi(n-1)}$ $u^{\pi(n-1)}$	$s - 2s_{\pi(n-1)}$ $u^{\pi(n-1)}$	π^*	p^*
		otherwise	BD(8)			

6.3   The $(n+1)$-Ray Variable Dimension Method Based on the D_{v1}-Triangulation

In this section we deal with how to use the D_{v1}-triangulation in the $(n+1)$-ray method proposed by Doup and Talman in [27]. Let $f : S^n \to R^{n+1}$ be continuous on S^n. Our purpose is to find a stationary point of f in S^n, i.e., a point x^* in S^n such that $(x - x^*)^\mathsf{T} f(x^*) \leq 0$ for all $x \in S^n$.

Let N_0 denote the index set $\{0, 1, \cdots, n\}$. For $i = 0, 1, \cdots, n$, let u^i denote the i-th unit vector in R^{n+1}. Take arbitrarily an interior point $x^0 \in S^n$ as an initial point. For $T \subset N_0$, let us define the set $A(T)$ by

$$A(T) = \left\{ x \in S^n \,\middle|\, \begin{array}{l} x = x^0 + \sum_{i \in T} \lambda_i (u^i - x^0), \\ \lambda_i \geq 0 \text{ for all } i \in T \end{array} \right\}$$

and the set $S^n(T)$ by

$$S^n(T) = \{x \in S^n \mid x_i = 0 \text{ for all } i \notin T\}.$$

It is obvious that the dimension of $A(T)$ is equal to t, where t is equal to $|T|$. Let H be a nonempty subset of N_0. Then the orthogonal projection vector of x^0 on $S^n(H)$, $v(H)$, is defined by

$$v_i(H) = \begin{cases} 0 & \text{if } i \notin H, \\ x_i^0 + (1 - \sum_{j \in H} x_j^0)/|H| & \text{if } i \in H, \end{cases}$$

for $i = 0, 1, \cdots, n$. When H is empty, it is conventional that $v(H) = x^0$. For a permutation $\gamma(T) = (\gamma(1), \gamma(2), \cdots, \gamma(t))$ of elements of T, let us define the set $A(\gamma(T))$ by

$$A(\gamma(T)) = \left\{ x \in S^n \,\middle|\, \begin{array}{l} x = x^0 + \sum_{j=1}^t \lambda_j w^j, \\ 1 \geq \lambda_1 \geq \lambda_2 \geq \cdots \geq \lambda_t \geq 0 \end{array} \right\},$$

where

$$w^1 = v(\{\gamma(1), \cdots, \gamma(t)\}) - x^0$$

and

$$w^j = v(\{\gamma(1); \cdots, \gamma(t - j + 1)\}) - v(\{\gamma(1), \cdots, \gamma(t - j + 2)\})$$

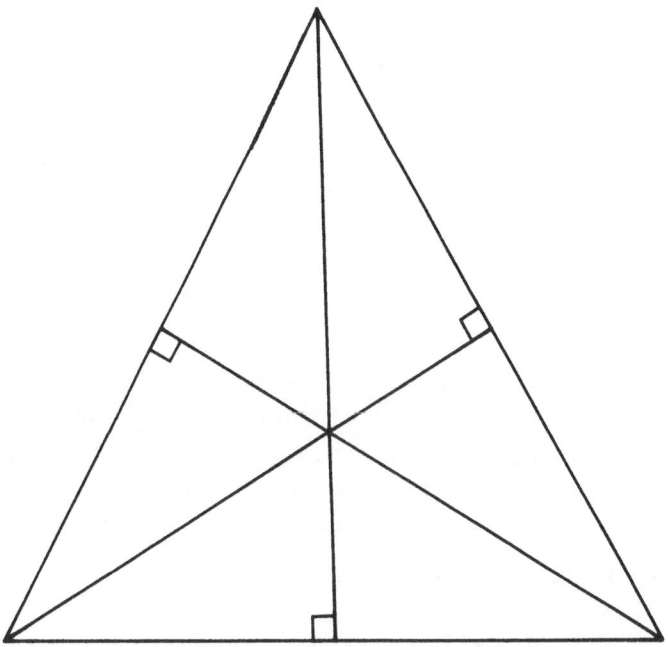

Figure 6.2: The sets $A(\gamma(T))$ in S^n for $n = 2$ and $x^0 = (1/3, 1/3, 1/3)^{\mathsf{T}}$.

for $j = 2, \cdots, t$. The sets $A(\gamma(T))$ in S^n, $T \subset N_0$, are illustrated in Figure 6.2 for $n = 2$ and $x^0 = (1/3, 1/3, 1/3)^{\mathsf{T}}$. Then it is clear that $A(\gamma(T))$ is t-dimensional and $A(T)$ is the union of $A(\gamma(T))$ over all permutations $\gamma(T)$ of the elements in T. Let W denote the $(n+1) \times t$ matrix with the i-th column equal to w^i for $i = 1, 2, \cdots, t$. Then

$$A(\gamma(T)) = \left\{ x^0 + m^{-1} W x \mid x \in C^t(m) \right\}.$$

This means that $A(\gamma(T))$ is homeomorphic to $C^t(m)$. Thus the D_{v1}-triangulation of $C^t(m)$ induces a simplicial subdivision of $A(\gamma(T))$ after the same transformation. It is denoted by $D_{v1}(\gamma(T))$. It can easily be seen that the union of $D_{v1}(\gamma(T))$ over all permutations of the elements in T is a simplicial subdivision of $A(T)$. It is written as $D_{v1}(T)$. There-fore, we obtain a simplicial subdivision of S^n with grid size m^{-1} by taking the union of $D_{v1}(T)$ over all $T \subset N_0$ with $|T| = n$.

For a nonempty $T \subseteq N_0$ with $t = |T|$, a k-simplex σ with vertices $z^0, z^1, \cdots, z^k$, for $k = t-1$ or t, is called T-complete if the system of

$n + 2$ linear equations

$$\sum_{i=0}^{k} \lambda_i \begin{pmatrix} f(z^i) \\ 1 \end{pmatrix} + \sum_{j \notin T} \mu_j \begin{pmatrix} u^j \\ 0 \end{pmatrix} - \beta \begin{pmatrix} u \\ 0 \end{pmatrix} = \begin{pmatrix} 0 \\ 1 \end{pmatrix}$$

has a solution $(\lambda, \mu, \beta)^\top$ such that $\lambda_i \geq 0$ for $i = 0, 1, \cdots, k$, and $\mu_j \geq 0$ for $j \notin T$, where u is the $(n + 1)$-vector with all components equal to one. A T-complete simplex σ is complete if for every $x \in \sigma$, $x_j = 0$ for all $j \notin T$.

Starting at x^0, the $(n+1)$-ray variable dimension method generates a sequence of adjacent T-complete t-simplices in $A(T)$ for varying sets T until a complete simplex is found. Such a simplex yields an approximate stationary point of f on S^n. As follows, we describe the steps of the algorithm in case the simplicial subdivision of each set $A(\gamma(T))$ is based on the D_{v1}-triangulation with some given grid size m^{-1}.

The $(n+1)$-Ray Variable Dimension Method Based on the D_{v1}-Triangulation:

Initialization: Let k denote the unique index such that

$$f_k(x^0) = \max_{0 \leq j \leq n} f_j(x^0).$$

Set $T = \{k\}$, $t = 1$, and $\gamma(T) = (k)$. Set $y = 0$, $\pi = (1)$, $s = 1$, and $p = 0$. Further, set $z^0 = x^0$ and $\tau_0 = \{z^0\}$. Finally, set $r = 0$.

Step 1: Let σ_r be the simplex in $D_{v1}(\gamma(T))$ corresponding to the simplex

$$D_{v1}(y, \pi, s, p).$$

Thus τ_r is a facet of σ_r. Let z^+ denote the vertex of σ_r opposite to τ_r. Perform a linear programming step with $(f(z^+), 1)^\top$ in the system of linear equations

$$\sum_{j=0}^{t-1} \lambda_j \begin{pmatrix} f(z^j) \\ 1 \end{pmatrix} + \sum_{j \notin T} \mu_j \begin{pmatrix} u^j \\ 0 \end{pmatrix} - \beta \begin{pmatrix} u \\ 0 \end{pmatrix} = \begin{pmatrix} 0 \\ 1 \end{pmatrix}.$$

If some μ_c becomes zero then set $z^t = z^+$ and go to **Step 2**; otherwise, some λ_d becomes zero, then set $z^- = z^d$ and $z^d = z^+$, and go to **Step 3**.

Step 2: If $N_0 = T \cup \{c\}$, then σ_r is complete and the algorithm terminates; otherwise, perform the following increasing dimension procedure. Set $\tau_{r+1} = \sigma_r$ and $r = r + 1$. Set $T = T \cup \{c\}$,

$$\gamma(T) = (\gamma(1), \cdots, \gamma(t), c),$$

$y_i = y_{i-1}$ and $s_i = s_{i-1}$ for $i = t+1, \cdots, 2$, $y_1 = y_2$ and $s_1 = s_2$, set $p = p + 1$ if $\pi^{-1}(1) \leq p - 1$, set

$$\pi = (\pi(1)+1, \cdots, \pi(\pi^{-1}(1)-1)+1, 2, 1, \pi(\pi^{-1}(1)+1)+1, \cdots, \pi(t)+1)$$

in case $s_1 = 1$, and set

$$\pi = (\pi(1)+1, \cdots, \pi(\pi^{-1}(1)-1)+1, 1, 2, \pi(\pi^{-1}(1)+1)+1, \cdots, \pi(t)+1)$$

in case $s_1 = -1$. Finally, set $t = t + 1$ and go to **Step 1**.

Step 3: Let y^i be the vertex of $D_{v1}(y, \pi, s, p)$ corresponding to the vertex z^-. Set τ_{r+1} equal to the facet of σ_r opposite to the vertex z^- and set $r = r + 1$. Consider **Table 6.2.1**. If one of the cases $BD(j)$, $j = 1, 2, \cdots, 8$, occurs, then τ_r lies in the boundary of $A(\gamma(T))$.

1. When one of the cases $BD(1)$, $BD(2)$ and $y_{\pi(1)} = m$, $BD(3)$ and $y_{\pi(i)} = m$, $BD(6)$, or $BD(8)$ occurs, then τ_r is a complete simplex and the algorithm terminates.

2. When one of the cases both $BD(2)$ and $y_{\pi(1)} = 0$ or both $BD(3)$ and $y_{\pi(i)} = 0$ occurs, set

$$\gamma(T) = (\gamma(2), \gamma(1), \cdots, \gamma(t)),$$

and go to **Step 1**.

3. When one of the case both $BD(4)$ and either $\pi(1) = 1$ and $s_1 = 1$ or $\pi(i) = 2$ and $s_2 = -1$ or the case both $BD(7)$ and either $\pi(1) = 1$ and $s_1 = 1$ or $\pi(i) = 2$ and $s_2 = -1$ occurs, then set $\kappa = \gamma(t)$, $T = T \setminus \{\gamma(t)\}$, $y_j = y_{j+1}$ and $s_j = s_{j+1}$ for $j = 2, \cdots, t - 1$, $\pi(j) = \pi(j) - 1$ if $\pi(j) \neq 1$ or 2 for $j < \max \{\pi^{-1}(1), \pi^{-1}(2)\}$,

$$\pi(\min \{\pi^{-1}(1), \pi^{-1}(2)\}) = 1,$$

$\pi(j-1) = \pi(j) - 1$ for $\max\{\pi^{-1}(1), \pi^{-1}(2)\} < j \le t$, $t = t - 1$, and go to **Step 4**.

4. When the case both BD(5) and either $\pi(i) = 1$ or $\pi(i+1) = 1$ occurs, then set $\kappa = \gamma(t)$, $T = T \setminus \{\gamma(t)\}$, $y_j = y_{j+1}$ and $s_j = s_{j+1}$ for $j = 2, \cdots, t-1$, $\pi(j) = \pi(j) - 1$ if $\pi(j) \neq 1$ or 2 for $j < \max\{\pi^{-1}(1), \pi^{-1}(2)\}$,

$$\pi(\min\{\pi^{-1}(1), \pi^{-1}(2)\}) = 1,$$

$\pi(j-1) = \pi(j) - 1$ for $\max\{\pi^{-1}(1), \pi^{-1}(2)\} < j \le t$, $p = p - 1$, $t = t - 1$, and go to **Step 4**.

5. When one of the cases 1) BD(4) and neither $\pi(i) = 1$ and $s_1 = 1$ nor $\pi(i) = 2$ and $s_2 = -1$ and 2) BD(7) and neither $\pi(i) = 1$ and $s_1 = 1$ nor $\pi(i) = 2$ and $s_2 = -1$ occurs, then set

$$\gamma(T) = (\gamma(1), \cdots, \gamma(t - \pi(i) + 2), \gamma(t - \pi(i) + 1), \cdots, \gamma(t))$$

if $s_{\pi(i)} = 1$, and

$$\gamma(T) = (\gamma(1), \cdots, \gamma(t - \pi(i) + 3), \gamma(t - \pi(i) + 2), \cdots, \gamma(t))$$

if $s_{\pi(i)} = -1$, and go to **Step 1**.

6. When BD(5) and both $\pi(i) \neq 1$ and $\pi(i+1) \neq 1$ occurs, then set

$$\gamma(T) = (\gamma(1), \cdots, \gamma(t - \pi(i) + 3), \gamma(t - \pi(i) + 2), \cdots, \gamma(t))$$

if $s_{\pi(i)} = 1$, and

$$\gamma(T) = (\gamma(1), \cdots, \gamma(t - \pi(i) + 2), \gamma(t - \pi(i) + 1), \cdots, \gamma(t))$$

if $s_{\pi(i)} = -1$, and go to **Step 1**.

When none of the cases BD(j), $j = 1, 2, \cdots, 8$, occurs, set $y = \bar{y}$, $\pi = \bar{\pi}$, $s = \bar{s}$, and $p = \bar{p}$ according to **Table 6.2.1**, and go to **Step 1**.

Step 4: Set σ_r equal to τ_r and perform a linear programming step with $(u^\kappa, 0)^\top$ in the system of linear equations

$$\sum_{j=0}^{t} \lambda_j \begin{pmatrix} f(z^j) \\ 1 \end{pmatrix} + \sum_{j \notin T, j \neq \kappa} \mu_j \begin{pmatrix} u^j \\ 0 \end{pmatrix} - \beta \begin{pmatrix} u \\ 0 \end{pmatrix} = \begin{pmatrix} 0 \\ 1 \end{pmatrix}.$$

If some μ_c becomes zero then go to **Step 2**; otherwise, some λ_d becomes zero, then set $z^- = z^d$ and $z^i = z^{i+1}$ for $i = d, \cdots, t-1$, and go to **Step 3**.

Assuming nondegeneracy in each linear programming step, the algorithm terminates in **Step 2** or in Case 1 of **Step 3** within a finite number of iterations with some complete simplex σ. Let $z^0, z^1, \cdots, z^k$, be the vertices of σ and let $(\lambda^*, \mu^*, \beta^*)^\top$ be the corresponding solution of the linear system, then $x^1 = \sum_{j=0}^{k} \lambda_j^* z^j$ lies in σ since $\lambda_j^* \geq 0$ for all j and $\sum_{j=0}^{k} \lambda_j^* = 1$. The vector x^1 can be considered as an approximate stationary point of f on S^n in the sense that $(x - x^1)^\top \bar{f}(x^1) \leq 0$ for all x in S^n, where $\bar{f}(x^1) = \sum_{j=0}^{k} \lambda^* f(z^j)$ is the piecewise linear approximation of f at x^1 with respect to the D_{v1}-triangulation. If the accuracy of approximation at x^1 is not satisfactory, then the $(n+1)$-ray method can be restarted at x^1 for some larger m. We remark that the steps of the algorithm are similar in case x^0 (or x^1) lies in the boundary of S^n.

6.4 The $(2^{n+1} - 2)$-Ray Variable Dimension Method Based on the D_{v1}-Triangulation

In this section we consider how to use the D_{v1}-triangulation in the $(2^{n+1} - 2)$-ray variable dimension method proposed by Doup, van der Laan and Talman in [26]. This algorithm is able to find a stationary point of f on S^n only when f is a complementary function. Let $f : S^n \to R^{n+1}$ be continuous on S^n such that $x^\top f(x) = 0$ for all $x \in S^n$, i.e., f is a complementary function on S^n. Our purpose is to find a point x^* in S^n such that $f(x^*) \leq 0$. Let N_0 denote the index set $\{0, 1, \cdots, n\}$. For $i = 0, 1, \cdots, n$, let u^i denote the i-th unit vector in R^{n+1}. Take arbitrarily $x^0 \in S^n$ as an initial point. Let H be a nonempty proper subset of N_0 and let

$$S^n(H) = \{x \in S^n \mid x_j = 0 \text{ for all } j \notin H\}.$$

Set $H_0 = \{j \in H \mid x_j^0 = 0\}$. Then the projection vector of x^0 on $S^n(H)$, v(H), is defined by

$$v_i(H) = \begin{cases} 0 & \text{if } i \notin H, \\ (1 - \sum_{j \in H} x_j^0)/(\sum_{j \in H} x_j^0 + |H_0|) & \text{if } i \in H_0, \\ x_i^0(1 + |H_0|)/(\sum_{j \in H} x_j^0 + |H_0|) & \text{otherwise,} \end{cases}$$

for $i = 0, 1, \cdots, n$. When H is empty, we define $v(H) = x^0$.

Let $g = (g_0, g_1, \cdots, g_n)^\top$ be a sign vector such that $g_i \in \{-1, 0, +1\}$ for $i = 0, 1, \cdots, n$. Let $I^-(g) = \{i \in N_0 \mid g_i = -1\}$, $I^0(g) = \{i \in N_0 \mid g_i = 0\}$, and $I^+(g) = \{i \in N_0 \mid g_i = +1\}$. For a sign vector g such that both $I^+(g)$ and $\{i \in I^-(g) \mid x_i^0 > 0\}$ are nonempty, we define

$$A(g) = \left\{ x \in S^n \;\middle|\; \begin{array}{l} x_k/x_k^0 = 1 + \alpha \text{ if } k \in I^+(g) \text{ and } x_k^0 > 0, \\ \beta \le x_k/x_k^0 \le 1 + \alpha \text{ if } k \in I^0(g) \text{ and } x_k^0 > 0, \\ \beta = x_k/x_k^0 \text{ if } k \in I^-(g) \text{ and } x_k^0 > 0, \\ x_k = \alpha \text{ if } k \in I^+(g) \text{ and } x_k^0 = 0, \\ 0 \le x_k \le \alpha \text{ if } k \in I^0(g) \text{ and } x_k^0 = 0, \\ 0 = x_k \text{ if } k \in I^-(g) \text{ and } x_k^0 = 0, \\ \text{with } 0 \le \beta \le 1 \le 1 + \alpha \end{array} \right\}.$$

It is obvious that $A(g)$ is a t-dimensional polyhedron with $t = |I^0(g)| + 1$ and that S^n is equal to the union of $A(g)$ over all the sign vectors g such that $|I^0(g)| = n - 1$. In addition, let $k_0 = 0$ and let $\gamma(g) = (k_1, k_2, \cdots, k_{t-1})$ be a permutation of the $t - 1$ elements in $I^0(g)$. Then we define

$$A(g, \gamma(g)) = \left\{ x \in S^n \;\middle|\; \begin{array}{l} x = x^0 + \sum_{j=1}^t \lambda_j w^j \\ \text{with } 0 \le \lambda_t \le \cdots \le \lambda_1 \le 1 \end{array} \right\},$$

where $w^1 = v(I^+(g) \cup \{k_1, \cdots, k_{t-1}\}) - x^0$ and

$$w^j = v(I^+(g) \cup \{k_1, \cdots, k_{t-j}\}) - v(I^+(g) \cup \{k_1, \cdots, k_{t-j+1}\})$$

for $j = 2, \cdots, t$. Then it is clear that $A(g, \gamma(g))$ is t-dimensional and that $A(g)$ is the union of $A(g, \gamma(g))$ over all permutations $\gamma(g)$ of the elements in $I^0(g)$. Let P be the $(n+1) \times t$-matrix with the i-th column equal to w^i for $i = 1, 2, \cdots, t$. Then

$$A(g, \gamma(g)) = \left\{ x^0 + m^{-1} P x \mid x \in C^t(m) \right\}.$$

This means that $A(g, \gamma(g))$ is homeomorphic to $C^t(m)$. Thus the D_{v1}-triangulation induces a simplicial subdivision of $A(g, \gamma(g))$, which is denoted by $D_{v1}(g, \gamma(g))$. Moreover, as can easily be shown, the union of $D_{v1}(g, \gamma(g))$ over all permutations of the elements in $I^0(g)$ is a simplicial subdivision of $A(g)$. We write it as $D_{v1}(g)$. Finally, a simplicial subdivision of S^n with grid size m^{-1} is obtained by taking the union of $D_{v1}(g)$ over all sign vectors g such that both $I^+(g)$ and $\{i \in I^-(g) \mid x_i^0 > 0\}$ are nonempty.

For $x \in R^{n+1}$, let the sign vector of x be defined by

$$\text{sgn}(x) = (\text{sgn}_0(x), \text{sgn}_1(x), \cdots, \text{sgn}_n(x))^\mathsf{T},$$

where

$$\text{sgn}_i(x) = \begin{cases} -1 & \text{if } x_i < 0, \\ 0 & \text{if } x_i = 0, \\ +1 & \text{if } x_i = +1, \end{cases}$$

for $i = 0, 1, \cdots, n$.

For a given sign vector g with $t = |I^0(g)| + 1$, a k-simplex σ with vertices $z^0, z^1, \cdots, z^k$, for $k = t - 1$ or t, is called g-complete if the $(n + 2)$-system of linear equations

$$\sum_{j=0}^{k} \lambda_j \begin{pmatrix} f(z^j) \\ 1 \end{pmatrix} - \sum_{j \notin I^0(g)} \mu_j g_j \begin{pmatrix} u^j \\ 0 \end{pmatrix} = \begin{pmatrix} 0 \\ 1 \end{pmatrix}$$

has a nonnegative solution $(\lambda, \mu)^\mathsf{T}$.

Starting at x^0 with $g = \text{sgn}(f(x^0))$, the $(2^{n+1} - 2)$-ray variable dimension method generates a sequence of adjacent g-complete t-simplices in $A(g)$ with g-complete common facets for varying sign vectors g until an approximate solution has been found. Below we describe the steps of the algorithm in case the simplicial subdivision of each set $A(g, \gamma(g))$

is based on the D_{v1}-triangulation with some given grid size m^{-1}.

The $(2^{n+1} - 2)$-Ray Variable Dimension Method Based on the D_{v1}-Triangulation:

Initialization: Without loss of generality we assume that $f_i(x^0) \neq 0$ for all i. Set $g = \mathrm{sgn}(f(x^0))$ and $t = 1$. Set $y = 0$, $\pi = (1)$, $s = 1$ and $p = 0$. Further, set $z^0 = x^0$ and $\tau_0 = \{z^0\}$. Finally, set $r = 0$.

Setp 1: Let σ_r be the simplex in $D_{v1}(g, \gamma(g))$ corresponding to the simplex

$$D_{v1}(y, \pi, s, p).$$

Thus τ_r is a facet of σ_r. Let z^+ denote the vertex of σ_r opposite to τ_r. Perform a linear programming step with $(f(z^+), 1)^\top$ in the system of linear equations

$$\sum_{j=0}^{t-1} \lambda_j \begin{pmatrix} f(z^j) \\ 1 \end{pmatrix} - \sum_{j \notin I^0(g)} \mu_j g_j \begin{pmatrix} u^j \\ 0 \end{pmatrix} = \begin{pmatrix} 0 \\ 1 \end{pmatrix}.$$

If some μ_c becomes zero, then set $z^t = z^+$ and go to **Step 2**; otherwise, some λ_d becomes zero, then set $z^- = z^d$ and $z^d = z^+$, and go to **Step 3**.

Step 2: When $I^+(g) = \{c\}$ or $\{j \in I^-(g) \mid x_j^0 > 0\} = \{c\}$, σ_r yields an approximate solution and the algorithm terminates; otherwise perform the following increasing dimension procedure. Set $\tau_{r+1} = \sigma_r$ and $r = r + 1$. When $c \in I^+(g)$, set $g_c = 0$,

$$\gamma(g) = (c, k_1, \cdots, k_{t-1}),$$

$y_{t+1} = 0$, $s_{t+1} = 1$, if $p = 0$ then set

$$\pi = (t + 1, \pi(1), \cdots, \pi(t)),$$

and if $p \geq 1$ then set

$$\pi = (t + 1, \pi(1), \cdots, \pi(t))$$

and $p = p + 1$. When $c \in I^-(g)$, set $g_c = 0$,

$$\gamma(g) = (k_1, \cdots, k_{t-1}, c),$$

$y_i = y_{i-1}$ and $s_i = s_{i-1}$ for $i = t + 1, \cdots, 2$, $y_1 = y_2$ and $s_1 = s_2$, $p = p + 1$ if $\pi^{-1}(1) \leq p - 1$,

$$\pi = (\pi(1)+1, \cdots, \pi(\pi^{-1}(1)-1)+1, 2, 1, \pi(\pi^{-1}(1)+1)+1, \cdots, \pi(t)+1)'$$

in case $s_1 = 1$, and

$$\pi = (\pi(1)+1, \cdots, \pi(\pi^{-1}(1)-1)+1, 1, 2, \pi(\pi^{-1}(1)+1)+1, \cdots, \pi(t)+1)$$

in case $s_1 = -1$. Set $t = t + 1$, and go to **Step 1**.

Step 3: Let y^i be the vertex of $D_{v1}(y, \pi, s, p)$ corresponding to the vertex z^-. Set τ_{r+1} equal to the facet of σ_r opposite to the vertex z^- and $r = r + 1$. Consider **Table 1**. If one of the cases BD(j), $j = 1, 2, \cdots, 8$, occurs, then τ_r lies in the boundary of $A(g, \gamma(g))$.

1. When one of the cases BD(1), BD(2) and $y_{\pi(1)} = m$, BD(3) and $y_{\pi(i)} = m$, BD(6), or BD(8) occurs, then τ_r yields an approximate solution and the algorithm terminates.

2. When the case both BD(2) and $y_{\pi(1)} = 0$ occurs, then set $\kappa = k_1$, $g_{k_1} = +1$, $k_j = k_{j+1}$ for $j = 1, 2, \cdots, t - 2$,

$$\pi = (\pi(2), \pi(3), \cdots, \pi(t)),$$

$p = p - 1$, and $t = t - 1$, and go to **Step 4**.

3. When the case both BD(3) and $y_{\pi(i)} = 0$ occurs, then set $\kappa = k_1$, $g_{k_1} = +1$, $k_j = k_{j+1}$ for $j = 1, 2, \cdots, t - 2$,

$$\pi = (\pi(2), \pi(3), \cdots, \pi(t)),$$

and $t = t - 1$, and go to **Step 4**.

4. When either the case both BD(4) and either $\pi(i) = 1$ and $s_1 = 1$ or $\pi(i) = 2$ and $s_2 = -1$ or the case both BD(7) and either $\pi(i) = 1$ and $s_1 = 1$ or $\pi(i) = 2$ and $s_2 = -1$ occurs, then set $\kappa = k_{t-1}$, $g_{k_{t-1}} = -1$, $y_j = y_{j+1}$ and $s_j = s_{j+1}$

for $j = 2, \cdots, t - 1$, $\pi(j) = \pi(j) - 1$ if $\pi(j) \neq 1$ or 2 for $j < \max\{\pi^{-1}(1), \pi^{-1}(2)\}$,

$$\pi(\min\{\pi^{-1}(1), \pi^{-1}(2)\}) = 1,$$

$\pi(j - 1) = \pi(j) - 1$ for $\max\{\pi^{-1}(1), \pi^{-1}(2)\} < j \leq t$, and $t = t - 1$, and go to **Step 4**.

5. When the case both BD(5) and either $\pi(i) = 1$ or $\pi(i+1) = 1$ occurs, then set $\kappa = k_{t-1}$, $g_{k_{t-1}} = -1$, $y_j = y_{j+1}$ and $s_j = s_{j+1}$ for $j = 2, \cdots, t - 1$, $\pi(j) = \pi(j) - 1$ if $\pi(j) \neq 1$ or 2 for $j < \max\{\pi^{-1}(1), \pi^{-1}(2)\}$,

$$\pi(\min\{\pi^{-1}(1), \pi^{-1}(2)\}) = 1,$$

$\pi(j - 1) = \pi(j) - 1$ for $\max\{\pi^{-1}(1), \pi^{-1}(2)\} < j \leq t$, $p = p - 1$, and $t = t - 1$, and go to **Step 4**.

6. When one of the cases 1) BD(4) and neither $\pi(i) = 1$ and $s_1 = 1$ nor $\pi(i) = 2$ and $s_2 = -1$ and 2) BD(7) and neither $\pi(i) = 1$ and $s_1 = 1$ nor $\pi(i) = 2$ and $s_2 = -1$ occurs, then set

$$\gamma(g) = (k_1, \cdots, k_{t-\pi(i)+1}, k_{t-\pi(i)}, \cdots, k_{t-1})$$

if $s_{\pi(i)} = 1$, and

$$\gamma(g) = (k_1, \cdots, k_{t-\pi(i)+2}, k_{t-\pi(i)+1}, \cdots, k_{t-1})$$

if $s_{\pi(i)} = -1$, and go to **Step 1**.

7. When BD(5) and both $\pi(i) \neq 1$ and $\pi(i + 1) \neq 1$ occurs, then set

$$\gamma(g) = (k_1, \cdots, k_{t-\pi(i)+2}, k_{t-\pi(i)+1}, \cdots, k_{t-1})$$

if $s_{\pi(i)} = 1$, and

$$\gamma(g) = (k_1, \cdots, k_{t-\pi(i)+1}, k_{t-\pi(i)}, \cdots, k_{t-1})$$

if $s_{\pi(i)} = -1$, and go to **Step 1**.

When none of the cases $BD(j)$, $j = 1, 2, \cdots, 8$, occurs, then set $y = \bar{y}$, $\pi = \bar{\pi}$, $s = \bar{s}$, and $p = \bar{p}$ according to **Table 6.2.1**, and go to **Step 1**.

Step 4: Set σ_r equal to τ_r and perform a linear programming step with $-g_\kappa(u^\kappa, 0)^\mathsf{T}$ in the system of linear equations

$$\sum_{j=0}^{t} \lambda_j \begin{pmatrix} f(z^j) \\ 1 \end{pmatrix} - \sum_{j \notin I^0(g), j \neq \kappa} \mu_j g_j \begin{pmatrix} u^j \\ 0 \end{pmatrix} = \begin{pmatrix} 0 \\ 1 \end{pmatrix}.$$

If some μ_c becomes zero, then go to **Step 2**; otherwise, some λ_d becomes zero, then set $z^- = z^d$ and $z^i = z^{i+1}$ for $i = d, \cdots, t - 1$, and go to **Step 3**.

Assuming nondegeneracy in each linear programming step, the algorithm terminates within a finite number of iterations in **Step 2** or in Case 1 of **Step 3** with some simplex σ. Let $z^0, z^1, \cdots, z^k$, be the vertices of σ_k and let $(\lambda^*, \mu^*)^\mathsf{T}$ be the solution corresponding to the linear system, then $x^1 = \sum_{j=0}^{k} \lambda_j^* z^j$ lies in σ since $\lambda_j^* \geq 0$ for all j and $\sum_{j=0}^{k} \lambda_j^* = 1$. The vector x^1 can be considered as an approximate complementary point of f on S^n. If the accuracy of approximation is not satisfactory, then the $(2^{n+1} - 2)$-ray method can be restarted at x^1 for some larger m, in order to improve the accuracy, as has been shown in [26].

Chapter 7

The D_1-Triangulation in Variable Dimension Algorithms on the Euclidean Space

Since van der Laan and Talman initiated the $(n + 1)$-ray variable dimension algorithm on the unit simplex, a number of variable dimension algorithms on R^n for computing solutions of nonlinear equations have been developed, such as the $(n+1)$-ray variable dimension method proposed by van der Laan and Talman in [113], the $2n$-ray variable dimension method by van der Laan and Talman in [113], the 2^n-ray variable dimension method by Wright in [195], the 2-ray variable dimension method independently by Saigal in [155] and by Yamamoto in [196], and the $(3^n - 1)$-ray variable dimension method by Kojima and Yamamoto in [96]. The simplicial subdivisions that underlie these variable dimension algorithms are the A^*-triangulation, the K_1-triangulation, the J_1-triangulation, or the K'-triangulation of R^n. However, as has been seen in Chapter 4, the D_1-triangulation is superior to both the K_1-triangulation and the J_1-triangulation according to measures of efficiency such as the number of simplices in a unit cube, the diameter, and the average directional density. Since the efficiency of the K'-triangulation according to these measures is the same as the one of the K_1-triangulation, the D_1-triangulation is also superior to that triangulation. Concerning the A^*-triangulation, it can only be used in the $(n+1)$-ray variable dimension method. However, this algorithm is far from efficient, compared to the other variable dimension methods. We refer to van der Laan and Seelen [107] and van der Laan and Talman [116]. This also holds for the 2-ray variable dimension method. Therefore, it is interesting to incorporate the D_1-triangulation in the variable dimension algorithms other than the $(n+1)$-ray and the 2-ray methods. Since it is not straightforward to use the D_1-triangulation in variable dimension algorithms on R^n except the $2n$-ray and the 2-ray variable dimension methods, in this chapter we deal with how to adapt the D_1-triangulation in variable dimension algorithms on R^n. A new version of the D_1-triangulation is proposed. It induces similarly to the D_1-triangulation a simplicial subdivision of each of the subsets, into which a variable dimension algorithm subdivides R^n. One can directly use this version in both the 2^n-ray and the $(3^n - 1)$-ray variable dimension algorithms. It is expected that the cost of computation can be reduced through using the D_1-triangulation in these variable dimension algorithms. This chapter is organized as follows. Section 1 introduces the new version of the D_1-triangulation, called the D_{v2}-triangulation. Its pivot rules are given in Section 2. We describe how to incorporate the D_1-triangulation in the $2n$-ray variable dimension method in Section 3 and how to use the D_{v2}-triangulation in the 2^n-ray variable dimension method in Section 4. This chapter is based on Dang and Talman's [18].

7.1   The D_{v2}-Triangulation

Let N denote the index set $\{1, 2, \cdots, n\}$. Let

$$W^n = \{x \in R^n \mid x_1 \geq x_2, x_3, \cdots, x_n \geq 0\}$$

and

$$D = \{y \in W^n \mid \text{all components of y are even}\}.$$

The new version of the D_1-triangulation, called the D_{v2}-triangulation, subdivides W^n into n-dimensional simplices.

Take $y \in D$. Let $I(y)$ and $J(y)$ denote the sets

$$I(y) = \{i \in N \mid y_1 = y_i\} \text{ and } J(y) = \{j \in N \mid y_1 > y_j\}.$$

Take a sign vector $s = (s_1, s_2, \cdots, s_n)^\top$ such that for all $i \in N$, if $y_i = 0$ then $s_i = 1$, and for all $i \in I(y)$, if $s_1 = -1$ then $s_i = -1$. It is obvious that if there exists $i \in I(y)$ with $s_i = 1$ then $s_1 = 1$. Let

$$K(y, s) = \{i \in I(y) \mid s_i = 1\}.$$

Let l denote the number of elements in $I(y)$ and h the number of elements in $K(y, s)$. Take an integer p such that when $h = 0$, if $l = n$ then $p = 0$ or $2 \le p \le n - 1$ and if $l < n$ then $0 \le p \le n - 1$, and when $h > 0$, if $h = n$ then $p = 0$ and if $h < n$ then $0 \le p \le n - 1$. Take a permutation $\pi = (\pi(1), \pi(2), \cdots, \pi(n))$ of the elements of N such that for r with $\pi(r) = 1$, if $h = 0$ then $j > r$ for all $j \ne r$ with $\pi(j) \in I(y)$ and if $h > 0$ then $j < r$ for all $j \ne r$ with $\pi(j) \in K(y, s)$, and when $h = 0$, if $p \ge 1$ then

$$\{\pi(k) \mid p \le k \le n\} \ne I(y)$$

and when $h > 0$, if $p \ge 1$ then

$$\{\pi(k) \mid p \le k \le n\} \ne \{\pi(k) \in K(y, s) \mid p \le k \le n\}.$$

When $h = 0$, let

$$g_i(1) = \begin{cases} -1 & \text{if } i \in I(y), \\ 0 & \text{otherwise,} \end{cases}$$

for $i = 1, 2, \cdots, n$, and for $j = 2, 3, \cdots, n$, let

$$g_i(j) = \begin{cases} s_i & \text{if } i = j, \\ 0 & \text{otherwise,} \end{cases}$$

for $i = 1, 2, \cdots, n$. When $h > 0$, for $j = 1, 2, \cdots, n$, if $\pi(j) \in K(y, s)$, let

$$g_i(\pi(j)) = \begin{cases} 1 & \text{if } i \in K(y, s) \text{ and } j \le \pi^{-1}(i), \\ 0 & \text{otherwise,} \end{cases}$$

for $i = 1, 2, \cdots, n$; otherwise, let

$$g_i(\pi(j)) = \begin{cases} s_{\pi(j)} & \text{if } i = \pi(j), \\ 0 & \text{otherwise,} \end{cases}$$

for $i = 1, 2, \cdots, n$. Finally, let u^i be the i-th unit vector in R^n for $i = 1, 2, \cdots, n$.

Definition 7.1.1. For the vector y, the permutation π, the sign vector s, and the number p given as above, the vectors y^0, y^1, $\cdots$, y^n are given as follows.

If $p = 0$, then $y^0 = y$ and

$$y^k = y + g(\pi(k)), \quad k = 1, 2, \cdots, n.$$

If $p \geq 1$, then $y^0 = y + s$,

$$y^k = y^{k-1} - s_{\pi(k)} u^{\pi(k)}, \quad k = 1, 2, \cdots, p - 1,$$
$$y^k = y + g(\pi(k)), \quad k = p, \cdots, n.$$

Let y^0, y^1, $\cdots$, y^n be obtained from the above definition. Then it is obvious that they are affinely independent. Thus their convex hull is a simplex with vertices y^0, y^1, $\cdots$, y^n. Let us denote this simplex by $D_{v2}(y, \pi, s, p)$. Let D_{v2} be the set of simplices $D_{v2}(y, \pi, s, p)$ for all y, π, s and p given as above. We shall show that D_{v2} is a triangulation of W^n.

For y, π, s and p given as above, let

$$q = | K(y, s) \cap \{\pi(k) \mid 1 \leq k < p\} |$$

and let

$$\{\pi(k) \in K(y, s) \mid 1 \leq k \leq n\}$$
$$= \{\pi(i_1), \pi(i_2), \cdots, \pi(i_h) \mid i_1 < i_2 < \cdots < i_h\}.$$

When $\pi(i) \in K(y, s)$ and $i > i_{q+1}$, then i_{-1} denotes the index i_{k-1} where k is such that $i = i_k$.

Lemma 7.1.2. The union of all σ in D_{v2} is equal to W^n.

Proof. Clearly, every simplex in D_{v2} lies in W^n. Let $x \in W^n$ be arbitrary. Then $x \in D_{v2}(y, \pi, s, p)$ with y, π, s and p determined as follows. Take the vector y equal to

$$y_i = \begin{cases} \lfloor x_i \rfloor & \text{if } \lfloor x_i \rfloor \text{ is even,} \\ \lfloor x_i \rfloor + 1 & \text{otherwise,} \end{cases}$$

for $i = 1, 2, \cdots, n$, and take the sign vector s equal to

$$s_i = \begin{cases} 1 & \text{if } \lfloor x_i \rfloor \text{ is even,} \\ -1 & \text{otherwise,} \end{cases}$$

for $i = 1, 2, \cdots, n$. It is obvious that $y \in D$.

Case 1: $h = 0$.

When $l = 1$, the proof is the same as that of **Lemma 4.1.3**. Suppose $l > 1$. Let

$$\mu_1 = s_1(x_1 - y_1),$$

$$\mu_i = s_i(x_i - y_i) - s_1(x_1 - y_1) \text{ for } i \in I(y) \backslash \{1\},$$

and

$$\mu_i = s_i(x_i - y_i) \text{ for } i \in J(y).$$

Let $\mu = \sum_{j=1}^n \mu_j$.

When $\mu \leq 1$, take $\pi(i) = i$ for $i = 1, 2, \cdots, n$, and $p = 0$. Let $\beta_0 = 1 - \mu$ and $\beta_i = \mu_i$ for $i = 1, 2, \cdots, n$. Then it is obvious that $\beta_k \geq 0$ for all k,

$$x = \sum_{j=0}^n \beta_j y^j \text{ and } \sum_{j=0}^n \beta_j = 1,$$

where y^j is obtained from **Definition 7.1.1** for $j = 0, 1, \cdots, n$. Thus

$$x \in D_{v2}(y, \pi, s, p).$$

Suppose $\mu > 1$. Choose π such that

$$s_{\pi(1)}(x_{\pi(1)} - y_{\pi(1)}) \leq s_{\pi(2)}(x_{\pi(2)} - y_{\pi(2)}) \leq \cdots \leq s_{\pi(n)}(x_{\pi(n)} - y_{\pi(n)})$$

and for r with $\pi(r) = 1$, $j > r$ for all $j \neq r$ with $\pi(j) \in I(y)$. Let p_{max} denote the largest $1 \leq p \leq n - 1$ such that

$$\{\pi(k) \mid p \leq k \leq n\} \neq I(y).$$

We show that we can take the integer p between 2 and p_{max} if $l = n$ and between 1 and p_{max} if $l < n$ such that the following system has a nonnegative solution,

$$
\begin{aligned}
\beta_0 &= s_{\pi(1)}\big(x_{\pi(1)} - y_{\pi(1)}\big), \\
\beta_1 &= s_{\pi(2)}\big(x_{\pi(2)} - y_{\pi(2)}\big) - s_{\pi(1)}\big(x_{\pi(1)} - y_{\pi(1)}\big), \\
&\cdots, \\
\beta_{p-2} &= s_{\pi(p-1)}\big(x_{\pi(p-1)} - y_{\pi(p-1)}\big) - s_{\pi(p-2)}\big(x_{\pi(p-2)} - y_{\pi(p-2)}\big), \\
\beta_{p-1} &= -s_{\pi(p-1)}\big(x_{\pi(p-1)} - y_{\pi(p-1)}\big) + \lambda_p, \\
\beta_k &= s_{\pi(k)}\big(x_{\pi(k)} - y_{\pi(k)}\big) - \alpha_k,
\end{aligned}
$$

for $k = p, p+1, \cdots, n$, where

$$
\lambda_p =
\begin{cases}
\big(\sum_{j=p}^{n} s_{\pi(j)}\big(x_{\pi(j)} - y_{\pi(j)}\big) - 1\big)/(n-p) \\
\qquad \text{if } r < p, \\
\big(\sum_{j=p}^{n} s_{\pi(j)}\big(x_{\pi(j)} - y_{\pi(j)}\big) - 1 - (l-1)s_1(x_1 - y_1)\big)/(n-p-l+1) \\
\qquad \text{otherwise,}
\end{cases}
$$

and

$$
\alpha_k =
\begin{cases}
\lambda_p & \text{if } r < p \text{ or both } r \geq p \text{ and } \pi(k) \notin I(y)\backslash\{1\}, \\
s_1(x_1 - y_1) & \text{if } r \geq p \text{ and } \pi(k) \in I(y)\backslash\{1\},
\end{cases}
$$

for $k = p, p+1, \cdots, n$. For $p = p_{max}$, if $\beta_{p-1} \geq 0$, it is obvious that $\beta_k \geq 0$ for all k, and we take $p = p_{max}$. Suppose $\beta_{p-1} < 0$ for $p = p_{max}$. Since $\mu > 1$, there exist $2 \leq p_0 \leq p_{max} - 1$ in case $l = n$, and $1 \leq p_0 \leq p_{max} - 1$ in case $l < n$ such that

$$
0 \leq -s_{\pi(p_0-1)}\big(x_{\pi(p_0-1)} - y_{\pi(p_0-1)}\big) + \lambda_{p_0}
$$

and either both $r = p_0 + 1$ and $p_0 = n - l$ or

$$
0 > -s_{\pi(p_0)}\big(x_{\pi(p_0)} - y_{\pi(p_0)}\big) + \lambda_{p_0+1},
$$

and thus,

$$
s_{\pi(p_0)}\big(x_{\pi(p_0)} - y_{\pi(p_0)}\big) - \alpha_{p_0} \geq 0.
$$

Hence, when we choose $p = p_0$, then $\beta_k \geq 0$ for all k. It is clear that

$$
x = \sum_{j=0}^{n} \beta_j y^j \text{ and } \sum_{j=0}^{n} \beta_j = 1,
$$

where y^j is obtained from **Definition 7.1.1** for $j = 0, 1, \cdots, n$. Thus

$$x \in D_{v2}(y, \pi, s, p).$$

Case 2: $h > 0$.

When $h = 1$, the proof is the same as that of **Lemma 4.1.3**.

When $h = n$, choose $p = 0$ and π such that

$$s_{\pi(1)}(x_{\pi(1)} - y_{\pi(1)}) \leq \cdots \leq s_{\pi(n)}(x_{\pi(n)} - y_{\pi(n)})$$

and $\pi(n) = 1$. Let

$$
\begin{aligned}
\beta_0 &= 1 - s_{\pi(n)}(x_{\pi(n)} - y_{\pi(n)}), \\
\beta_1 &= s_{\pi(1)}(x_{\pi(1)} - y_{\pi(1)}), \\
\beta_2 &= s_{\pi(2)}(x_{\pi(2)} - y_{\pi(2)}) - s_{\pi(1)}(x_{\pi(1)} - y_{\pi(1)}), \\
&\cdots, \\
\beta_n &= s_{\pi(n)}(x_{\pi(n)} - y_{\pi(n)}) - s_{\pi(n-1)}(x_{\pi(n-1)} - y_{\pi(n-1)}).
\end{aligned}
$$

It is obvious that $\beta_k \geq 0$ for all k,

$$x = \sum_{j=0}^{n} \beta_j y^j \text{ and } \sum_{j=0}^{n} \beta_j = 1,$$

where y^j is obtained from **Definition 7.1.1** for $j = 0, 1, \cdots, n$. Thus

$$x \in D_{v2}(y, \pi, s, p).$$

Suppose $1 < h < n$. Let $\mu_1 = s_1(x_1 - y_1)$ and $\mu_i = s_i(x_i - y_i)$ for $i \in N \backslash K(y, s)$. Let $\mu = \mu_1 + \sum_{i \in N \backslash K(y,s)} \mu_i$. Choose π such that

$$s_{\pi(1)}(x_{\pi(1)} - y_{\pi(1)}) \leq \cdots \leq s_{\pi(n)}(x_{\pi(n)} - y_{\pi(n)})$$

and for r with $\pi(r) = 1$, $j < r$ for all $j \neq r$ with $\pi(j) \in K(y, s)$.

Suppose $\mu \leq 1$. Take $p = 0$. Let $\beta_0 = 1 - \mu$ and

$$
\beta_i = \begin{cases}
s_{\pi(i)}(x_{\pi(i)} - y_{\pi(i)}) - s_{\pi(i-1)}(x_{\pi(i-1)} - y_{\pi(i-1)}) \\
\qquad\qquad \text{if } \pi(i) \in K(y, s) \text{ and } i > i_1, \\
s_{\pi(i)}(x_{\pi(i)} - y_{\pi(i)}) \qquad \text{otherwise},
\end{cases}
$$

for $i = 1, 2, \cdots, n$. Then it is obvious that $\beta_k \geq 0$ for all k,

$$x = \sum_{j=0}^{n} \beta_j y^j \text{ and } \sum_{j=0}^{n} \beta_j = 1,$$

where y^j is obtained from **Definition 7.1.1** for $j = 0, 1, \cdots, n$. Thus

$$x \in D_{v2}(y, \pi, s, p).$$

Suppose $\mu > 1$. Let p_{max} denote the largest $1 \leq p \leq n - 1$ such that

$$\{\pi(k) \mid p \leq k \leq n\} \neq \{\pi(k) \in K(y, s) \mid p \leq k \leq n\}.$$

We show that we can take the integer p, $1 \leq p \leq p_{max}$, such that the following system has a nonnegative solution,

$$
\begin{aligned}
\beta_0 &= s_{\pi(1)}(x_{\pi(1)} - y_{\pi(1)}), \\
\beta_1 &= s_{\pi(2)}(x_{\pi(2)} - y_{\pi(2)}) - s_{\pi(1)}(x_{\pi(1)} - y_{\pi(1)}), \\
&\cdots, \\
\beta_{p-2} &= s_{\pi(p-1)}(x_{\pi(p-1)} - y_{\pi(p-1)}) - s_{\pi(p-2)}(x_{\pi(p-2)} - y_{\pi(p-2)}), \\
\beta_{p-1} &= -s_{\pi(p-1)}(x_{\pi(p-1)} - y_{\pi(p-1)}) + c(p), \\
\beta_i &= s_{\pi(i)}(x_{\pi(i)} - y_{\pi(i)}) - \nu_i,
\end{aligned}
$$

for $i = p, p + 1, \cdots, n$, where

$$
c(p) = \begin{cases} (\sum_{j=p}^{n} s_{\pi(j)}(x_{\pi(j)} - y_{\pi(j)}) - 1)/(n - p) & \text{if } r < p, \\ (\sum_{j=p}^{n} \rho_{\pi(j)} - 1)/(n - p - h + q + 1) & \text{otherwise} \end{cases}
$$

with

$$
\rho_{\pi(j)} = \begin{cases} 0 & \text{if } \pi(j) \in K(y, s) \text{ and } i_{q+1} \leq j < i_h, \\ s_{\pi(j)}(x_{\pi(j)} - y_{\pi(j)}) & \text{otherwise,} \end{cases}
$$

for $j = p, p + 1, \cdots, n$, and where

$$
\nu_i = \begin{cases} s_{\pi(i-1)}(x_{\pi(i-1)} - y_{\pi(i-1)}) & \text{if } i > i_{q+1} \text{ and } \pi(i) \in K(y, s), \\ c(p) & \text{otherwise,} \end{cases}
$$

for $i = p, p + 1, \cdots, n$. For $p = p_{max}$, if $\beta_{p-1} \geq 0$, it is obvious that $\beta_k \geq 0$ for all k, and we take $p = p_{max}$. Otherwise, since $\mu > 1$, there exists $1 \leq p_0 \leq p_{max} - 1$ such that

$$0 \leq -s_{\pi(p_0-1)}(x_{\pi(p_0-1)} - y_{\pi(p_0-1)}) + c(p_0)$$

and

$$0 > -s_{\pi(p_0)}(x_{\pi(p_0)} - y_{\pi(p_0)}) + c(p_0 + 1),$$

and hence,

$$s_{\pi(p_0)}(x_{\pi(p_0)} - y_{\pi(p_0)}) - \nu_{p_0} \geq 0.$$

Thus when we take $p = p_0$, $\beta_k \geq 0$ for all k. Obviously,

$$x = \sum_{j=0}^{n} \beta_j y^j \text{ and } \sum_{j=0}^{n} \beta_j = 1,$$

where y^j is obtained from **Definition 7.1.1** for $j = 0, 1, \cdots, n$. Thus

$$x \in D_{v2}(y, \pi, s, p).$$

From the above conclusions, the lemma follows immediately.
END

Lemma 7.1.3. For σ^1 and σ^2 in D_{v2}, the intersection of σ^1 and σ^2 is either a common face of both σ^1 and σ^2 or empty.

proof. The proof is similar to the one of **Lemma 4.1.4**.
END

Theorem 7.1.4. D_{v2} is a triangulation of W^n.

Proof. From **Definition 7.1.1**, **Lemma 7.1.2** and **Lemma 7.1.3**, the theorem follows immediately.
END

This simplicial subdivision is called the D_{v2}-triangulation of W^n. The D_{v2}-triangulation is illustrated in Figure 7.1 for $n = 3$ and $x_1 \leq 2$.

7.2 Pivot Rules of the D_{v2}-Triangulation

Let $\sigma = D_{v2}(y, \pi, s, p)$ be a simplex of the D_{v2}-triangulation with vertices y^0, y^1, $\cdots$, y^n. We want to obtain the parameters of the simplex $\bar{\sigma} = D_{v2}(\bar{y}, \bar{\pi}, \bar{s}, \bar{p})$ such that all vertices of σ are also vertices of $\bar{\sigma}$ except

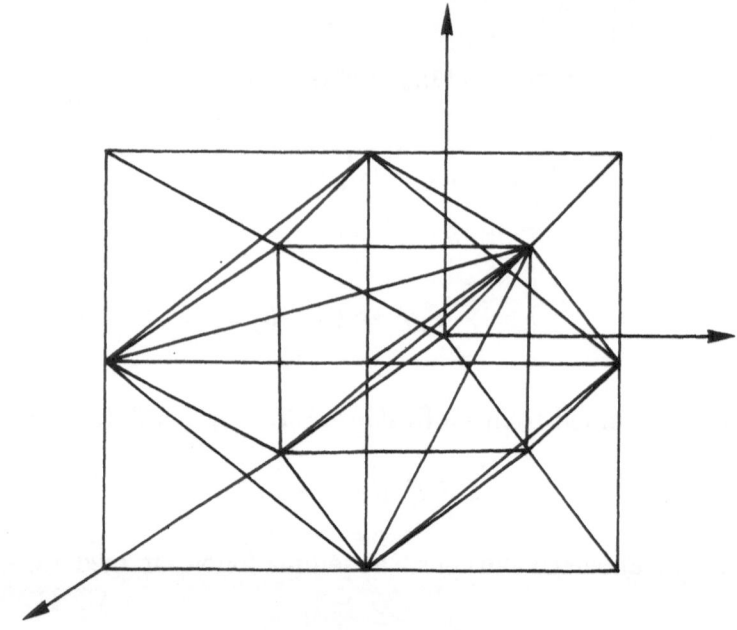

Figure 7.1: The D_{v2}-triangulation of W^n for $n = 3$ and $x_1 \leq 2$.

the vertex y^i, in case the facet of σ opposite to the vertex y^i does not lie in the boundary of W^n. In **Table 7.2.1**, we show how $\bar{y}$, $\bar{\pi}$, $\bar{s}$ and $\bar{p}$ depend on y, π, s, p and i. From this table, it is easy to obtain each vertex $\bar{y}^k$ of $\bar{\sigma}$, $k = 0, 1, \cdots, n$, and in particular its vertex opposite to the facet shared with σ.

In this table, j^* is equal to $\pi(k)$ with k such that $k \neq i$, $\pi(k) \neq 1$ and $n - 2 \leq k \leq n$. Moreover, if $\pi(n-1) = 1$, $s_1 = -1$ and $\pi(n-1), \pi(n) \in I(\bar{y})$ then $p^\# = p - 1$, and if $\pi(n-1), \pi(n) \in K(\bar{y}, \bar{s})$ then

$$
p^\# = \begin{cases}
k & \text{if there exists } 1 \leq k \leq n - 2 \text{ satisfying } \pi(k) \notin K(\bar{y}, \bar{s}) \text{ and} \\
& \quad \pi(j) \in K(\bar{y}, \bar{s}) \text{ for } k < j \leq n, \\
0 & \text{otherwise;}
\end{cases}
$$

otherwise, $p^\# = p$. Finally, if $\pi(n) = 1$, $s_1 = -1$ and $\pi(n-1), \pi(n) \in I(\bar{y})$ then

$$
\pi^* = (\pi(1), \pi(2), \cdots, \pi(n-2), \pi(n), \pi(n-1))
$$

and $p^* = p - 1$, and if $\pi(n-1), \pi(n) \in K(\bar{y}, \bar{s})$ then

$$
\pi^* = (\pi(1), \pi(2), \cdots, \pi(n-2), \pi(n), \pi(n-1))
$$

and

$$p^* = \begin{cases} k & \text{if there exists } 1 \le k \le n-2 \text{ satisfying } \pi(k) \notin K(\bar{y}, \bar{s}) \text{ and} \\ & \pi(j) \in K(\bar{y}, \bar{s}) \text{ for } k < j \le n, \\ 0 & \text{otherwise;} \end{cases}$$

otherwise, $\pi^* = \pi$ and $p^* = p$.

Table 7.2.1(1). The Pivot Rules of the D_{v2}-Triangulation for $n \geq 2$

i	p	h	condition	condition	$\bar{y}$	$\bar{s}$	$\bar{\pi}$	$\bar{p}$
0	0	$h=0$	$l=n$	$n=2$	$y+2s_2u^2$	$s-2s_2u^2$	π	1
				$n \geq 3$	y	s	π	2
			$l<n$		y	s	π	1
		$h>0$	$h=n$		$y+2s_1u^1$	$s-2s_1u^1$	π	$n-1$
			$h<n$		y	s	π	1
	1				y	s	π	0
	$p \geq 2$			$y_{\pi(1)} \neq 0$	y	$s-2s_{\pi(1)}u^{\pi(1)}$	π	p
				$y_{\pi(1)} = 0$	case(1)		π	p
$1 \leq i < n$	0	$h=0$	$\pi(i) \in J(y)$ or $\pi(i)=1$	$y_{\pi(i)} \neq 0$	y	$s-2s_{\pi(i)}u^{\pi(i)}$	π	p
				$y_{\pi(i)} = 0$	case(2)			
			$\pi(i) \in I(y)$, $\pi(i) \neq 1$		BD(1)			
		$h>0$	$\pi(i) \in K(y,s)$, $i_1 < i < \pi^{-1}(1)$		y	s	$(\pi(1),\ldots,\pi(i_{-1}-1),\ \pi(i),\pi(i_{-1}+1),\ldots,\pi(i-1),\ \pi(i_{-1}),\pi(i+1),\ldots,\pi(n))$	p
			$i_1 < i = \pi^{-1}(1)$					
			$i=i_1$ or $\pi(i) \notin K(y,s)$	$y_{\pi(i)} \neq 0$	BD(2)	$s-'2s_{\pi(i)}u^{\pi(i)}$	$(\pi(i),\pi(1),\ldots,\pi(i-1),\ \pi(i+1),\ldots,\pi(n))$	p
				$y_{\pi(i)} = 0$	case(3)			
$1 \leq i < p-1$		$h=0$	$\pi(i)=1$	$\pi(i+1) \in I(y)$	BD(3)	s	$(\pi(1),\ldots,\pi(i+1),\ \pi(i),\ldots,\pi(n))$	p
				$\pi(i+1) \notin I(y)$	y		$(\pi(1),\ldots,\pi(i+1),\ \pi(i),\ldots,\pi(n))$	
			$\pi(i) \neq 1$		y	s	$(\pi(1),\ldots,\pi(i+1),\ \pi(i),\ldots,\pi(n))$	
		$h>0$	$\pi(i) \in K(y,s)$		BD(4)	s	$(\pi(1),\ldots,\pi(i+1),\ \pi(i),\ldots,\pi(n))$	p
			$\pi(i) \notin K(y,s)$		y	s	$(\pi(1),\ldots,\pi(i+1),\ \pi(i),\ldots,\pi(n))$	p
$i = p-1$	$p \geq 2$	$h=0$	$\pi(i)=1$, $\pi(k) \in I(y)$ for all $p < k \leq n$		y	s	π	$p-2$
			otherwise		y	s	π	$p-1$
		$h>0$			y	s	π	$p-1$

Table 7.2.1(2). The Pivot Rules of the D_{v2}-Triangulation for $n \geq 2$

i	p	h			$\bar{y}$	$\bar{s}$	$\bar{\pi}$	$\bar{p}$
$p \leq i \leq n$	$1 \leq p$, $< n-1$	$h = 0$	$\pi^{-1}(1) < p$		y	s	$(\pi(1),\ldots,\pi(p-1),$ $\pi(i),\pi(p),\ldots,\pi(i-1),$ $\pi(i+1),\ldots,\pi(n))$	$p+1$
			$\pi(i) \in I(y)$ $\pi^{-1}(1) \geq p$	$\pi(i) \neq 1$ $\pi(i) = 1$	BD(5) y	s	$(\pi(1),\ldots,\pi(p-1),\pi(i),$ $\pi(p),\ldots,\pi(i-1),$ $\pi(i+1),\ldots,\pi(n))$	$p+1$
			$\pi^{-1}(1) \geq p$ $\pi(i) \notin I(y)$	$\pi(j) \in I(y)$ for $p \leq j \leq n$ $(j \neq i)$, $p = n-2$	$y + 2s_{j^*} u^{j^*}$	$s - 2s_{j^*} u^{j^*}$	$(\pi(1),\ldots,\pi(p-1),\pi(i),$ $\pi(p),\ldots,\pi(i-1),$ $\pi(i+1),\ldots,\pi(n))$	$p+1$
				$\pi(j) \in I(y)$ for $p \leq j \leq n$ $(j \neq i)$, $p < n-2$	y	s	$(\pi(1),\ldots,\pi(p-1),\pi(i),$ $\pi(p),\ldots,\pi(i-1),$ $\pi(i+1),\ldots,\pi(n))$	$p+2$
				not all $p \leq j \leq n$ $(j \neq i)$ satisfy $\pi(j) \in I(y)$.	y	s	$(\pi(1),\ldots,\pi(p-1),\pi(i),$ $\pi(p),\ldots,\pi(i-1),$ $\pi(i+1),\ldots,\pi(n))$	$p+1$
		$h > 0$	$\pi(i) \in K(y,s)$	$i_{q+1} < i$ $< \pi^{-1}(1)$	y	s	$(\pi(1),\ldots,\pi(i-1-1),\pi(i),$ $\pi(i-1+1),\ldots,\pi(i-1),$ $\pi(i-1),\pi(i+1),\ldots,\pi(n))$	p
				$i_{q+1} < i = \pi^{-1}(1)$ $i = i_{q+1}$	BD(6) y	s	$(\pi(1),\ldots,\pi(p-1),\pi(i),$ $\pi(p),\ldots,\pi(i-1),$ $\pi(i+1),\ldots,\pi(n))$	$p+1$
			$\pi(i) \notin K(y,s)$	$\pi(j) \in K(y,s)$ for $p \leq j \leq n$ $(j \neq i)$	$y + 2s_1 u^1$	$s - 2s_1 u^1$	$(\pi(1),\ldots,\pi(p-1),\pi(i),$ $\pi(p),\ldots,\pi(i-1),$ $\pi(i+1),\ldots,\pi(n))$	$n-1$
				not all $p \leq j \leq n$ $(j \neq i)$ satisfy $\pi(j) \in K(y,s)$	y	s	$(\pi(1),\ldots,\pi(p-1),\pi(i),$ $\pi(p),\ldots,\pi(i-1),$ $\pi(i+1),\ldots,\pi(n))$	$p+1$
$i = n-1$	$1 \leq p$, $= n-1$				$y + 2s_{\pi(n)} u^{\pi(n)}$	$s - 2s_{\pi(n)} u^{\pi(n)}$	π	$p^{\#}$
$i = n$					$y + 2s_{\pi(n-1)} u^{\pi(n-1)}$, $u^{\pi(n-1)}$	$s - 2s_{\pi(n-1)}$, $u^{\pi(n-1)}$	π^{*}	p^{*}

7.3 The $2n$-Ray Variable Dimension Method Based on the D_1-Triangulation

In this section we deal with how to use the D_1-triangulation in the $2n$-ray variable dimension algorithm given by van der Laan and Talman in [113]. Similarly one can apply the D_1-triangulation to the 2-ray variable dimension method proposed independently by Saigal in [155] and by Yamamoto in [196]. Let $f : R^n \to R^n$ be continuous. We want to compute a zero point of f, i.e., a vector $x^* \in R^n$ such that $f(x^*) = 0$.

Take arbitrarily an initial point $x^0 \in R^n$. Let N denote the index set $\{1, 2, \cdots, n\}$. For a sign vector $t = (t_1, t_2, \cdots, t_n)^\top$ with $t_i \in \{-1, 0, +1\}$ for all $i \in N$ and $t \neq 0$, let us define the set

$$A(t) = \left\{ x \in R^n \mid x_i = x_i^0 - \lambda_i t_i, \ \lambda_i \geq 0 \text{ for all } i \in N \right\}.$$

It is clear that the dimension of $A(t)$ is equal to d, where d is the number of nonzero components of t, and that $A(t)$ is homeomorphic to the set

$$R_+^d = \left\{ x \in R^d \mid x_i \geq 0 \text{ for } i = 1, 2, \cdots, d \right\}.$$

Thus from the D_1-triangulation of R^d restricted to R_+^d, we obtain a simplicial subdivision of $A(t)$ with grid size equal to one. The triangulation of $A(t)$ is denoted by $D_1(t)$. By multiplying the simplices of the D_1-triangulation with δ, $\delta > 0$, we obtain a simplicial subdivision of $A(t)$ with grid size equal to δ.

Let σ denote a k-dimensional simplex in $D_1(t)$ with vertices v^0, v^1, $\cdots$, v^k, $k = d-1$ or d. The simplex σ is called t-complete if the system of linear equations

$$\sum_{j=0}^{k} \lambda_j \begin{pmatrix} f(v^j) \\ 1 \end{pmatrix} + \sum_{j \in N, \, t_j = 0} \mu_j \begin{pmatrix} u^j \\ 0 \end{pmatrix} - \beta \begin{pmatrix} t \\ 0 \end{pmatrix} = \begin{pmatrix} 0 \\ 1 \end{pmatrix}$$

has a solution $(\lambda, \mu, \beta)^\top$ such that $\lambda_j \geq 0$ for all $0 \leq j \leq k$ and $|\mu_j| < \beta$ for all j with $t_j = 0$. If we have that $\beta = 0$ at a solution $(\lambda, \mu, \beta)^\top$, then we call this simplex complete.

Starting at x^0, the $2n$-ray variable dimension algorithm on R^n generates a sequence of adjacent d-simplices in $A(t)$ having t-complete facets for varying sign vectors t until a complete simplex is found. Such a simplex yields an approximate zero point of f.

The $2n$-Ray Variable Dimension Method Based on the D_1-Triangulation:

Initialization: Without loss of generality we assume $f(x^0) \neq 0$ and there is a unique index r such that $|f_r(x^0)| = \max_{1 \leq i \leq n} |f_i(x^0)|$. Set $t_r = \text{sgn}(f_r(x^0))$, $t_i = 0$ for all $i \neq r$, and $z_1 = r$. Next, set $v^0 = x^0$ and $\tau_0 = \{v^0\}$. Further, set $y = 0$, $\pi = (1)$, $s = 1$ and $p = 0$. Finally, set $k = 0$ and $d = 1$.

Step 1: Let σ_k be the simplex in $D_1(t)$ corresponding to the simplex $D_1(y, \pi, s, p)$. Thus τ_k is a facet of σ_k. Let v^+ denote the vextex of σ_k opposite to τ_k. Perform a linear programming step with $(f(v^+), 1)^\top$ in the system of linear equations

$$\sum_{j=0}^{d-1} \lambda_j \begin{pmatrix} f(v^j) \\ 1 \end{pmatrix} + \sum_{j \in N, \, t_j = 0} \mu_j \begin{pmatrix} u^j \\ 0 \end{pmatrix} - \beta \begin{pmatrix} t \\ 0 \end{pmatrix} = \begin{pmatrix} 0 \\ 1 \end{pmatrix}$$

If some $|\mu_c|$ becomes equal to $\beta > 0$, then set $v^d = v^+$ and $t_c = -\text{sgn}(\mu_c)$, and go to **Step 2**; if some λ_g becomes zero, then set $v^- = v^g$ and $v^g = v^+$, and go to **Step 3**; if β becomes zero, then σ_k is a complete simplex and the algorithm terminates.

Step 2: Set $\tau_{k+1} = \sigma_k$ and perform the following increasing dimension procedure. Set $z_{d+1} = c$, $y_{d+1} = 0$,

$$\pi = (\pi(1), \cdots, \pi(d), d+1)$$

if $p = 0$,

$$\pi = (d+1, \pi(1), \cdots, \pi(d))$$

if $p \geq 1$, $s_{d+1} = 1$, and $p = p + 1$ if $p \geq 1$. Set $k = k + 1$ and $d = d + 1$, and go to **Step 1**.

Step 3: Let y^i be the vertex of $D_1(y, \pi, s, p)$ corresponding to the vertex v^-. Set τ_{k+1} equal to the facet of σ_k opposite to the vertex v^- and $k = k + 1$. Consider **Table 4.2.1**.

1. When $i \geq 1$, $y_{\pi(i)} = 0$, $s_{\pi(i)} = 1$ and $p = 0$, set $y_j = y_{j+1}$ and $s_j = s_{j+1}$ for $\pi(i) \leq j \leq d - 1$, $\pi(j) = \pi(j) - 1$ if both $j < i$ and $\pi(j) > \pi(i)$, and for $i < j \leq d$ set

$$\pi(j-1) = \begin{cases} \pi(j) - 1 & \text{if } \pi(j) > \pi(i), \\ \pi(j) & \text{otherwise.} \end{cases}$$

Set $\kappa = z_{\pi(i)}$, $t_\kappa = 0$, and $d = d - 1$, and go to **Step 4**.

2. When $i = 0$, $y_{\pi(1)} = 0$, $s_{\pi(1)} = 1$ and $p \geq 2$, set $y_j = y_{j+1}$ and $s_j = s_{j+1}$ for $\pi(1) \leq j \leq d - 1$, for $1 < j \leq d$ set

$$\pi(j-1) = \begin{cases} \pi(j) - 1 & \text{if } \pi(j) > \pi(1), \\ \pi(j) & \text{otherwise,} \end{cases}$$

and $p = p - 1$. Set $\kappa = z_{\pi(1)}$, $t_\kappa = 0$, and $d = d - 1$, and go to **Step 4**.

When the above cases do not occur, then set $y = \bar{y}$, $\pi = \bar{\pi}$, $s = \bar{s}$ and $p = \bar{p}$ according to **Table 4.2.1**, and go to **Step 1**.

Step 4: Set $\sigma_k = \tau_k$ and perform a linear programming step with $(u^\kappa, 0)^\mathsf{T}$ in the system of linear equations

$$\sum_{j=0}^{d} \lambda_j \begin{pmatrix} f(v^j) \\ 1 \end{pmatrix} + \sum_{j \in N,\, j \neq \kappa,\, t_j = 0} \mu_j \begin{pmatrix} u^j \\ 0 \end{pmatrix} - \beta \begin{pmatrix} t \\ 0 \end{pmatrix} = \begin{pmatrix} 0 \\ 1 \end{pmatrix}.$$

If some $|\mu_c|$ becomes equal to $\beta > 0$, then go to **Step 2**; otherwise, some λ_g becomes zero, then set $v^- = v^g$ and $v^j = v^{j+1}$ for $j = g, \cdots, d - 1$, and go to **Step 3**.

Assuming nondegeneracy, then under the convergence condition given in [113], the algorithm terminates in **Step 1** within a finite number of iterations with a d-dimensional complete simplex σ. Let v^0, v^1, $\cdots$, v^d be the vertices of σ and let $(\lambda^*, \mu^*, \beta^*)^\mathsf{T}$ be the corresponding solution of the linear system. Then $\beta^* = 0$, $\mu_j^* = 0$ for j with $t_j = 0$, $\sum_{j=0}^{d} \lambda_j^* = 1$, and $\lambda_j^* \geq 0$ for all j. We obtain that $\sum_{j=0}^{d} \lambda_j^* f(v^j) = 0$. The point $x^1 = \sum_{j=0}^{d} \lambda_j^* v^j$ can therefore be considered as an approximate zero point of f and lies in σ. If the accuracy of approximation is not satisfactory then the $2n$-ray variable dimension method can be restarted at x^1 with a finer simplicial subdivision in order to improve the accuracy of approximation.

7.4 The 2^n-Ray Variable Dimension Algorithm Based on the D_{v2}-Triangulation

In this section we consider how to incorporate the D_{v2}-triangulation in the 2^n-ray variable dimension method proposed by Wright in [195]. It can similarly be derived how to apply the D_{v2}-triangulation to the $(3^n - 1)$-ray variable dimension method proposed by Kojima and Yamamoto in [96].

Let $t = (t_1, t_2, \cdots, t_n)^\top$ denote a sign vector such that $t_i \in \{-1, 0, +1\}$ for $i = 1, 2, \cdots, n$. Let us define the set $H(t)$ by

$$H(t) = \{i \in N \mid t_i = 0\}.$$

Let T denote the set of all such sign vectors t with $t \neq 0$. Take arbitrarily an initial point $x^0 \in R^n$. For $t \in T$, let $X(t)$ denote the set

$$\left\{ x^0 + x \in R^n \;\middle|\; \begin{array}{l} t_i x_i = t_j x_j \geq 0 \text{ if } t_i \neq 0 \text{ and } t_j \neq 0 \\ t_i x_i \geq |x_j| \text{ if } t_i \neq 0 \text{ and } t_j = 0 \end{array} \right\}.$$

Then it is obvious that for $t^1, t^2 \in T$, $X(t^1) \cap X(t^2)$ is a common face of both $X(t^1)$ and $X(t^2)$ and that the union of $X(t)$ over all $t \in T$ is equal to R^n. Let d denote the dimension of $X(t)$ for $t \in T$, i.e., $d = |H(t)| + 1$.

The 2^n-ray variable dimension method is based on a simplicial subdivision of R^n which satisfies that its restriction on every subset $X(t)$ induces a simplicial subdivision of $X(t)$. For a given $t \in T$, we will derive a triangulation of $X(t)$ from the D_{v2}-triangulation of W^d. Let $Z(t)$ denote the sign vector set

$$\left\{ z = (z_1, z_2, \cdots, z_n)^\top \;\middle|\; \begin{array}{l} z_j = t_j \text{ for } j \in N \text{ with } t_j \neq 0 \\ z_j \in \{-1, +1\} \text{ for } j \in N \text{ with } t_j = 0 \end{array} \right\}.$$

For $z \in Z(t)$, let us define the set $X(t, z)$ by

$$X(t, z) = \left\{ x^0 + x \in R^n \;\middle|\; \begin{array}{l} z_i x_i = z_i x_j \geq 0 \text{ if } t_i \neq 0 \text{ and } t_j \neq 0 \\ z_i x_i \geq z_j x_j \geq 0 \text{ if } t_i \neq 0 \text{ and } t_j = 0 \end{array} \right\}.$$

It is obvious that the set $X(t)$ is equal to the union of $X(t, z)$ over all $z \in Z(t)$. Let $k_2, k_3, \cdots, k_d$ denote the indices such that $t_{k_j} = 0$ for

$j = 2, \cdots, d$. Next, let B denote the $n \times d$ matrix with the first column equal to the sign vector t and the j-th column equal to the n-vector $z_{k_j} u^{k_j}$ for $j = 2, \cdots, d$. Then it can easily be seen that

$$X(t, z) = \left\{ x^0 + Bx \mid x \in W^d \right\}.$$

Thus $X(t, z)$ is homeomorphic to the set W^d. Then given the D_{v2}-triangulation of W^d, we obtain a simplicial subdivision of $X(t, z)$ with grid size equal to one under the same transformation. It is denoted by $D_{v2}(X(t, z))$. Note that by multiplying the simplices of the D_{v2}-triangulation with $\delta, \delta > 0$, we obtain a simplicial subdivision of $X(t, z)$ with grid size equal to δ. Thus, using the D_{v2}-triangulation of W^d, we derive a simplicial subdivision of the subset $X(t)$ by taking the union of $D_{v2}(X(t, z))$ over all $z \in Z(t)$, which is represented by $D_{v2}(X(t))$. It is also obvious that the union of simplicial subdivisions of all subsets, $\cup_{t \in T} D_{v2}(X(t))$, induces a triangulation of R^n. Therefore, a simplicial subdivision for the 2^n-ray variable dimension method is obtained.

Let $f : R^n \to R^n$ be continuous. We want to compute a zero point of f, i.e., a vector $x^* \in R^n$ such that $f(x^*) = 0$.

Definition 7.4.1. A k-simplex σ with vertices $v^0, v^1, \cdots, v^k$ for $k = d - 1$ or d is t-complete if the system of linear equations

$$\sum_{i=0}^{k} \lambda_i \begin{pmatrix} f(v^i) \\ 1 \end{pmatrix} + \sum_{i \notin H(t)} \mu_i t_i \begin{pmatrix} u^i \\ 0 \end{pmatrix} = \begin{pmatrix} 0 \\ 1 \end{pmatrix}$$

has a nonnegative solution (λ, μ). If $\mu = 0$ at a nonnegative solution (λ, μ), then the simplex σ is complete.

Starting at x^0 with $t = -\text{sgn}(f(x^0))$, the 2^n-ray method generates for varying sign vectors t a sequence of adjacent d-dimensional simplices in $X(t)$ having t-complete common facets until a complete simplex is yielded. Then an approximate zero point of f has been found (see Wright [195]).

The 2^n-Ray Variable Dimension Method Based on the D_{v2}-Triangulation:

Initialization: Without loss of generality we assume that $f_i(x^0) \neq 0$

for all i. Set

$$t_i = \begin{cases} -1 & \text{if } f_i(x^0) > 0, \\ 1 & \text{if } f_i(x^0) < 0, \end{cases}$$

for $i = 1, 2, \cdots, n$. Next, set $z = t$, $y = 0$, $s = 1$, $\pi = (1)$ and $p = 0$. Further, set $v^0 = x^0$ and $\tau_0 = \{v^0\}$. Finally, set $k = 0$ and $d = 1$.

Step 1: Let σ_k be the simplex in $D_{v2}(X(t, z))$ corresponding to the simplex $D_{v2}(y, \pi, s, p)$. Thus τ_k is a facet of σ_k. Let v^+ denote the vertex of σ_k opposite to τ_k. Perform a linear programming step with $(f(v^+), 1)^\top$ in the system of linear equations

$$\sum_{l=0}^{d-1} \lambda_l \begin{pmatrix} f(v^l) \\ 1 \end{pmatrix} + \sum_{r \notin H(t)} \mu_r t_r \begin{pmatrix} u^r \\ 0 \end{pmatrix} = \begin{pmatrix} 0 \\ 1 \end{pmatrix}.$$

If some μ_b becomes zero then set $v^d = v^+$, and go to **Step 2**; otherwise, some λ_c becomes zero, then set $v^- = v^c$ and $v^c = v^+$, and go to **Step 3**.

Step 2: When $d = n$, σ_k is a complete simplex and the algorithm terminates. Otherwise perform the following increasing dimension procedure. Set $\tau_{k+1} = \sigma_k$ and $k = k + 1$. Next, set $t_c = 0$, $k_{d+1} = c$, $y_{d+1} = y_1$ and $s_{d+1} = s_1$. Let m denote the integer such that $\pi(m) = 1$. Then π and p are adapted as follows.

1. When $p = 0$, if $s_1 = -1$, then set

$$\pi = (\pi(1), \cdots, \pi(d), d + 1);$$

otherwise, set

$$\pi = (\pi(1), \cdots, \pi(m - 1), d + 1, \pi(m), \cdots, \pi(d)).$$

2. When $p \geq 1$, if $m < p$ and $s_1 = -1$, then set

$$\pi = (\pi(1), \cdots, \pi(m), d + 1, \pi(m + 1), \cdots, \pi(d))$$

and $p = p + 1$; if $m < p$ and $s_1 = 1$, then set

$$\pi = (\pi(1), \cdots, \pi(m - 1), d + 1, \pi(m), \cdots, \pi(d))$$

and $p = p + 1$; if $m \geq p$ and $s_1 = -1$, then set

$$\pi = (\pi(1), \cdots, \pi(d), d + 1);$$

if $m \geq p$ and $s_1 = 1$, then set

$$\pi = (\pi(1), \cdots, \pi(m-1), d+1, \pi(m), \cdots, \pi(d)).$$

Set $d = d + 1$, and go to **Step 1**.

Step 3: Let y^i be the vertex of $D_{v2}(y, \pi, s, p)$ corresponding to the vertex v^-. Consider **Table 7.2.1**. If one of the cases BD(1), BD(2), BD(3), BD(4), BD(5) or BD(6) occurs, then the facet τ_{k+1} of σ_k opposite to the vertex v^- lies in the boundary of $X(t)$.

1. When either BD(1) or BD(5) occurs, set $t_{k_{\pi(i)}} = z_{k_{\pi(i)}}$, $\kappa = k_{\pi(i)}$,

$$k_\iota = \begin{cases} k_\iota & \text{if } \iota < \pi(i), \\ k_{\iota+1} & \text{if } \pi(i) \leq \iota, \end{cases}$$

for $\iota = 1, 2, \cdots, d-1$,

$$y_\iota = \begin{cases} y_\iota & \text{if } \iota < \pi(i), \\ y_{\iota+1} & \text{if } \pi(i) \leq \iota, \end{cases}$$

for $\iota = 1, 2, \cdots, d-1$,

$$\pi(\iota) = \begin{cases} \pi(\iota) & \text{if } \pi(\iota) < \pi(i) \text{ and } \iota < i, \\ \pi(\iota+1) & \text{if } \pi(\iota+1) < \pi(i) \text{ and } i \leq \iota, \\ \pi(\iota) - 1 & \text{if } \pi(\iota) > \pi(i) \text{ and } \iota < i, \\ \pi(\iota+1) - 1 & \text{if } \pi(\iota+1) > \pi(i) \text{ and } i \leq \iota, \end{cases}$$

for $\iota = 1, 2, \cdots, d-1$, and

$$s_\iota = \begin{cases} s_\iota & \text{if } \iota < \pi(i), \\ s_{\iota+1} & \text{if } \pi(i) \leq \iota, \end{cases}$$

for $\iota = 1, 2, \cdots, d-1$. Finally, set σ_{k+1} equal to τ_{k+1}, $d = d-1$ and $k = k+1$, and go to **Step 4**.

2. When either BD(2) or BD(6) occurs, set $t_{k_{\pi(i_{-1})}} = z_{k_{\pi(i_{-1})}}$, $\kappa = k_{\pi(i_{-1})}$,

$$k_\iota = \begin{cases} k_\iota & \text{if } \iota < \pi(i_{-1}), \\ k_{\iota+1} & \text{if } \pi(i_{-1}) \leq \iota, \end{cases}$$

for $\iota = 1, 2, \cdots, d-1$,

$$y_\iota = \begin{cases} y_\iota & \text{if } \iota < \pi(i_{-1}), \\ y_{\iota+1} & \text{if } \pi(i_{-1}) \leq \iota, \end{cases}$$

for $\iota = 1, 2, \cdots, d-1$,

$$\pi(\iota) = \begin{cases} \pi(\iota) & \text{if } \pi(\iota) < \pi(i_{-1}) \text{ and } \iota < i_{-1}, \\ \pi(\iota+1) & \text{if } \pi(\iota+1) < \pi(i_{-1}) \text{ and } i_{-1} \leq \iota, \\ \pi(\iota) - 1 & \text{if } \pi(\iota) > \pi(i_{-1}) \text{ and } \iota < i_{-1}, \\ \pi(\iota+1) - 1 & \text{if } \pi(\iota+1) > \pi(i_{-1}) \text{ and } i_{-1} \leq \iota, \end{cases}$$

for $\iota = 1, 2, \cdots, d-1$, and

$$s_\iota = \begin{cases} s_\iota & \text{if } \iota < \pi(i_{-1}), \\ s_{\iota+1} & \text{if } \pi(i_{-1}) \leq \iota, \end{cases}$$

for $\iota = 1, 2, \cdots, d-1$. Finally, set σ_{k+1} equal to τ_{k+1}, $d = d - 1$ and $k = k + 1$, and go to **Step 4**.

3. When BD(3) occurs, set $t_{k_{\pi(i+1)}} = z_{k_{\pi(i+1)}}$, $\kappa = k_{\pi(i+1)}$,

$$k_\iota = \begin{cases} k_\iota & \text{if } \iota < \pi(i+1), \\ k_{\iota+1} & \text{if } \pi(i+1) \leq \iota, \end{cases}$$

for $\iota = 1, 2, \cdots, d-1$,

$$y_\iota = \begin{cases} y_\iota & \text{if } \iota < \pi(i+1), \\ y_{\iota+1} & \text{if } \pi(i+1) \leq \iota, \end{cases}$$

for $\iota = 1, 2, \cdots, d-1$,

$$\pi(\iota) = \begin{cases} \pi(\iota) & \text{if } \pi(\iota) < \pi(i+1) \text{ and } \iota < i+1, \\ \pi(\iota+1) & \text{if } \pi(\iota+1) < \pi(i+1) \text{ and } i+1 \leq \iota, \\ \pi(\iota) - 1 & \text{if } \pi(\iota) > \pi(i+1) \text{ and } \iota < i+1, \\ \pi(\iota+1) - 1 & \text{if } \pi(\iota+1) > \pi(i+1) \text{ and } i+1 \leq \iota, \end{cases}$$

for $\iota = 1, 2, \cdots, d-1$,

$$s_\iota = \begin{cases} s_\iota & \text{if } \iota < \pi(i+1), \\ s_{\iota+1} & \text{if } \pi(i+1) \leq \iota, \end{cases}$$

for $\iota = 1, 2, \cdots, d-1$, and $p = p-1$. Finally, set σ_{k+1} equal to τ_{k+1}, $d = d-1$ and $k = k+1$, and go to **Step 4**.

4. When BD(4) occurs, set $t_{k_{\pi(i)}} = z_{k_{\pi(i)}}$, $\kappa = k_{\pi(i)}$,

$$k_\iota = \begin{cases} k_\iota & \text{if } \iota < \pi(i), \\ k_{\iota+1} & \text{if } \pi(i) \leq \iota, \end{cases}$$

for $\iota = 1, 2, \cdots, d-1$,

$$y_\iota = \begin{cases} y_\iota & \text{if } \iota < \pi(i), \\ y_{\iota+1} & \text{if } \pi(i) \leq \iota, \end{cases}$$

for $\iota = 1, 2, \cdots, d-1$,

$$\pi(\iota) = \begin{cases} \pi(\iota) & \text{if } \pi(\iota) < \pi(i) \text{ and } \iota < i, \\ \pi(\iota+1) & \text{if } \pi(\iota+1) < \pi(i) \text{ and } i \leq \iota, \\ \pi(\iota)-1 & \text{if } \pi(\iota) > \pi(i) \text{ and } \iota < i, \\ \pi(\iota+1)-1 & \text{if } \pi(\iota+1) > \pi(i) \text{ and } i \leq \iota, \end{cases}$$

for $\iota = 1, 2, \cdots, d-1$,

$$s_\iota = \begin{cases} s_\iota & \text{if } \iota < \pi(i), \\ s_{\iota+1} & \text{if } \pi(i) \leq \iota, \end{cases}$$

for $\iota = 1, 2, \cdots, d-1$, and $p = p-1$. Finally, set σ_{k+1} equal to τ_{k+1}, $d = d-1$ and $k = k+1$, and go to **Step 4**.

5. In all other cases in **Table 7.2.1** the facet τ_{k+1} of σ_k opposite to the vertex v^- does not lie in the boundary of $X(t)$. If case(1) occurs then set $z_{k_{\pi(1)}} = -z_{k_{\pi(1)}}$; if either case(2) or case(3) occurs then set $z_{k_{\pi(i)}} = -z_{k_{\pi(i)}}$; otherwise set $y = \bar{y}$, $\pi = \bar{\pi}$, $s = \bar{s}$ and $p = \bar{p}$ according to **Table 7.2.1**. Set $k = k+1$, and go to **Step 1**.

Step 4: Perform a linear programming step with $t_\kappa(u^\kappa, 0)^\mathsf{T}$ in the system of linear equations

$$\sum_{l=0}^{d} \lambda_l \begin{pmatrix} f(v^l) \\ 1 \end{pmatrix} + \sum_{r \notin H(t), r \neq \kappa} \mu_r t_r \begin{pmatrix} u^r \\ 0 \end{pmatrix} = \begin{pmatrix} 0 \\ 1 \end{pmatrix}.$$

If some μ_b becomes zero, then go to **Step 2**; otherwise, some λ_c becomes zero, set $v^- = v^c$ and $v^l = v^{l+1}$ for $l = c, \cdots, d-1$, and go to **Step 3**.

Assuming nondegeneracy, then under the convergence condition given in [195], the algorithm terminates in **Step 2** within a finite number of iterations. Then an n-dimensional simplex σ is obtained with vertices $v^0, v^1, \cdots, v^n$ for which $\sum_{j=0}^{n} \lambda_j^* f(v^j) = 0$, $\sum_{j=0}^{n} \lambda_j^* = 1$, and $\lambda_j^* \geq 0$ for all j, where λ^* is the corresponding solution of the system. The point $x^1 = \sum_{j=0}^{n} \lambda_j^* v^j$ can be considered as an approximate zero point of f and lies in σ. If the accuracy of approximation is not satisfactory then the 2^n-ray variable dimension method can be restarted at x^1 with a finer simplicial subdivision in order to improve the accuracy of approximation.

Chapter 8

The D_3-Triangulation for Simplicial Homotopy Algorithms

Simplicial homotopy algorithms on the unit simplex S^n were initiated by Eaves in [34] for computing fixed points. Eaves's algorithm is built upon a homotopy function and starts at a complete simplex on the trivial level. It generates a sequence of adjacent $(n+1)$-dimensional complete simplices with varying grid sizes until a complete simplex yields a satisfactorily approximate solution. The idea of homotopies was generalized by Eaves and Saigal to R^n in [42]. Triangulations of continuous refinement of grid sizes play a basic role in simplicial homotopy algorithms. In order to develop more efficient simplicial homotopy methods, a number of triangulations of continuous refinement of grid sizes have been proposed such as the K'_3-triangulation for $(0,1] \times S^n$ of Eaves in [34], the K_3-triangulation for $(0,1] \times R^n$ of Eaves and Saigal in [42], the J'_3-triangulation for $(0,1] \times S^n$ and the J_3-triangulation for $(0,1] \times R^n$ of Todd in [177], the triangulations for $(0,1] \times S^n$ and $(0,1] \times R^n$ of van der Laan and Talman in [111], the triangulations for $(0,1] \times S^n$ and $(0,1] \times R^n$ of Shamir in [166], the triangulations for $(0,1] \times R^n$ of Kojima and Yamamoto in [95], the triangulations for $(0,1] \times R^n$ of Broadie and Eaves in [5] and the triangulations for $(0,1] \times \Pi_{j=1}^m S^{n_j}$ of Doup and Talman in [29]. All these triangulations are based on either the K_1-triangulation or the J_1-triangulation of R^n.

As was seen in Chapter 4, the D_1-triangulation is superior to both
the K_1-triangulation and the J_1-triangulation of R^n according to measures of efficiency such as the number of simplices in the unit cube,
the diameter and the average directional density. So it is significant to
construct triangulations of continuous refinement of grid sizes based on
the D_1-triangulation. In this chapter we focus on triangulations of continuous refinement of grid sizes for $(0,1] \times R^n$ with a fixed refinement
factor of two. Based on the D_1-triangulation, a new triangulation of
continuous refinement of grid sizes for $(0,1] \times R^n$ is proposed having
a fixed refinement factor of two. This simplicial subdivision is called
the D_3-triangulation and is such that on each level 2^{-k}, $k = 0, 1, \cdots$, it
triangulates R^n according to the D_1-triangulation with grid size 2^{-k}. It
is shown that the D_3-triangulation is superior to the K_3-triangulation
and the J_3-triangulation for $(0,1] \times R^n$ according to measures of efficiency. This chapter is organized as follows. Section 1 introduces the
simplices of the D_3-triangulation. In Section 2 we prove that the collection of these simplices indeed yields a triangulation of continuous grid
refinement. Its pivot rules are described in Section 3. We compare the
D_3-triangulation with the K_3-triangulation and the J_3-triangulation in
Section 4. This chapter is based on Dang's [14].

8.1   Definition of the D_3-Triangulation

In this section we give an algebraic definition of the new triangulation
for simplicial homotopy algorithms, based on the D_1-triangulation of
R^n. Let u^i denote the i-th unit vector in R^{n+1} for $i = 0, 1, \cdots, n$. Next,
let N_0 denote the index set $\{0, 1, \cdots, n\}$ and let D denote the set

$$\left\{ y \in (0,1] \times R^n \;\middle|\; \begin{array}{l} y = (y_0, y_1, \cdots, y_n)^\top, \text{ and} \\ \text{for some integer } k \geq 0, \; y_0 = 2^{-(k+1)} \text{ and} \\ y_i/y_0 \text{ is odd for } i = 1, 2, \cdots, n \end{array} \right\}.$$

Take $y \in D$. Let

$$t_i = \begin{cases} -1 & \text{if } y_i/y_0 = 1 (\mathrm{mod} 4), \\ 1 & \text{if } y_i/y_0 = 3 (\mathrm{mod} 4), \end{cases}$$

for $i = 1, 2, \cdots, n$. Take a permutation $\pi = (\pi(0), \pi(1), \cdots, \pi(n))$ of the elements of N_0. Let j denote the integer such that $\pi(j) = 0$. Take a sign vector $s = (s_0, s_1, \cdots, s_n)^\mathsf{T}$ such that $s_0 = -1$, $s_i \in \{-1, 1\}$ for $i = 1, 2, \cdots, n$ and $s_{\pi(i)} = t_{\pi(i)}$ for $i = j + 1, \cdots, n$. Next, let

$$w_{\pi(i)} = (y_{\pi(i)} + y_0 s_{\pi(i)})/y_0$$

for $i = 0, 1, \cdots, j - 1$ and let I denote the set

$$\left\{ \pi(i) \mid w_{\pi(i)}/2 \text{ is even and } 0 \le i \le j - 1 \right\}.$$

In addition, let h denote the number of elements in the set I. Finally, take p_1 and p_2 to be integers such that $-1 \le p_1 \le j - 2$ and if $h = 0$ then $0 \le p_2 \le n - j - 1$ and if $h > 0$ then $p_2 = n - j$.

Definition 8.1.1. Let y, π, s, p_1 and p_2 be given as above. Then y^{-1}, $y^0, \cdots, y^n$ are given as follows.

If $p_1 = -1$, then $y^{-1} = y$ and

$$y^i = y + y_0 s_{\pi(i)} u^{\pi(i)}, \quad i = 0, 1, \cdots, j - 1.$$

If $p_1 \ge 0$, then

$$
\begin{aligned}
y^{-1} &= y + y_0 \sum_{l=0}^{j-1} s_{\pi(l)} u^{\pi(l)}, \\
y^i &= y^{i-1} - y_0 s_{\pi(i)} u^{\pi(i)}, \quad i = 0, 1, \cdots, p_1 - 1, \\
y^i &= y + y_0 s_{\pi(i)} u^{\pi(i)}, \quad i = p_1, \cdots, j - 1.
\end{aligned}
$$

If $h > 0$, then

$$
\begin{aligned}
y^j &= y + y_0 \sum_{l=0}^{j-1} s_{\pi(l)} u^{\pi(l)} + y_0 \sum_{l=j+1}^{n} s_{\pi(l)} u^{\pi(l)} + y_0 u^0, \\
y^i &= y^{i-1} - 2 y_0 s_{\pi(i)} u^{\pi(i)}, \quad i = j + 1, \cdots, n.
\end{aligned}
$$

If $h = 0$ and $p_2 = 0$, then

$$
\begin{aligned}
y^j &= y + y_0 \sum_{l=0}^{j-1} s_{\pi(l)} u^{\pi(l)} - y_0 \sum_{l=j+1}^{n} s_{\pi(l)} u^{\pi(l)} + y_0 u^0, \\
y^i &= y^j + 2 y_0 s_{\pi(i)} u^{\pi(i)}, \quad i = j + 1, \cdots, n.
\end{aligned}
$$

Finally, if $h = 0$ and $p_2 \ge 1$, then

$$
\begin{aligned}
y^j &= y + y_0 \sum_{l=0}^{j-1} s_{\pi(l)} u^{\pi(l)} + y_0 \sum_{l=j+1}^{n} s_{\pi(l)} u^{\pi(l)} + y_0 u^0, \\
y^i &= y^{i-1} - 2 y_0 s_{\pi(i)} u^{\pi(i)}, \quad i = j + 1, \cdots, j + p_2 - 1, \\
y^i &= y^* + 2 y_0 s_{\pi(i)} u^{\pi(i)}, \quad i = j + p_2, \cdots, n,
\end{aligned}
$$

where

$$y^* = y + y_0 \sum_{l=0}^{j-1} s_{\pi(l)} u^{\pi(l)} - y_0 \sum_{l=j+1}^{n} s_{\pi(l)} u^{\pi(l)} + y_0 u^0.$$

Let y^{-1}, y^0, $\cdots$, y^n be obtained from **Definition 8.1.1**. Then it is obvious that y^{-1}, y^0, $\cdots$, y^n are affinely independent. Thus the convex hull of y^{-1}, y^0, $\cdots$, y^n is a simplex. We write this simplex as $D_3(y, \pi, s, p_1, p_2)$. Let D_3 denote the set of simplices $D_3(y, \pi, s, p_1, p_2)$ for y, π, s, p_1 and p_2 given as above. We show that D_3 is a triangulation of continuous refinement of grid sizes of $(0, 1] \times R^n$ by a constructing process in the following section.

8.2   Construction of the D_3-triangulation

In this section we show that the set D_3 defined in Section 1 is a triangulation with a fixed refinement factor of two for $(0, 1] \times R^n$. Let N denote the index set $\{1, 2, \cdots, n\}$ and let Q denote the set of all vectors in R^n whose components are integers. Take an arbitrary element w in Q. Then we define

$$I_o(w) = \{i \in N \mid w_i \text{ is odd}\} \text{ and } I_e(w) = \{j \in N \mid w_j \text{ is even}\}.$$

Furthermore, let $A(w)$ denote the set

$$\{x \in R^n \mid w_i - 1 \leq x_i \leq w_i + 1 \text{ for } i \in I_o(w) \text{ and } x_i = w_i \text{ for } i \in I_e(w)\}$$

and let $B(w)$ denote the set

$$\{x \in R^n \mid x_i = w_i \text{ for } i \in I_o(w) \text{ and } w_i - 1 \leq x_i \leq w_i + 1 \text{ for } i \in I_e(w)\}.$$

Let k be a nonnegative integer. Finally, let $D^k(w)$ denote the convex hull of the set

$$\left(\left\{2^{-k}\right\} \times A(w)\right) \cup \left(\left\{2^{-(k+1)}\right\} \times B(w)\right).$$

Lemma 8.2.1. The set $D^k(w)$ is equal to the set C defined by

$$C = \left\{ d \in [2^{-(k+1)}, 2^{-k}] \times R^n \ \middle| \ \begin{array}{l} |d_i - w_i| \leq 2^{k+1} d_0 - 1 \text{ for } i \in I_o(w) \\ |d_i - w_i| \leq 2 - 2^{k+1} d_0 \text{ for } i \in I_e(w) \end{array} \right\}.$$

Proof. Take $\bar{x} = (x_0, x)^\top \in D^k(w)$. Then there exist

$$\bar{x}^A = (2^{-k}, x^A)^\top \in \left\{2^{-k}\right\} \times A(w),$$

$$\bar{x}^B = (2^{-(k+1)}, x^B)^\top \in \left\{2^{-(k+1)}\right\} \times B(w),$$

$\lambda_A \geq 0$ and $\lambda_B \geq 0$ such that

$$\lambda_A + \lambda_B = 1 \text{ and } \bar{x} = \lambda_A \bar{x}^A + \lambda_B \bar{x}^B.$$

Thus

$$x_i - w_i = (2^{k+1}x_0 - 1)(x_i^A - w_i) \text{ for } i \in I_o(w)$$

and

$$x_i - w_i = (2 - 2^{k+1}x_0)(x_i^B - w_i) \text{ for } i \in I_e(w).$$

Therefore, $\bar{x} \in C$. This means that $D^k(w) \subseteq C$.

Next, take $\bar{x} = (x_0, x)^\top \in C$, $\lambda_A = 2^{k+1}x_0 - 1$ and $\lambda_B = 2 - 2^{k+1}x_0$. If $\lambda_A = 0$ or $\lambda_B = 0$, then, respectively,

$$\bar{x} \in \left\{2^{-(k+1)}\right\} \times B(w) \text{ or } \bar{x} \in \left\{2^{-k}\right\} \times A(w).$$

When $\lambda_A > 0$ and $\lambda_B > 0$, choose x^A and x^B such that

$$x_i^A = (x_i - (2 - 2^{k+1}x_0)w_i)/(2^{k+1}x_0 - 1) \text{ and } x_i^B = w_i \text{ for } i \in I_o(w)$$

and

$$x_i^A = w_i \text{ and } x_i^B = (x_i - (2^{k+1}x_0 - 1)w_i)/(2 - 2^{k+1}x_0) \text{ for } i \in I_e(w).$$

Let $\bar{x}^A = (2^{-k}, x^A)^\top$ and $\bar{x}^B = (2^{-(k+1)}, x^B)^\top$. Then

$$\bar{x}^A \in \left\{2^{-k}\right\} \times A(w) \text{ and } \bar{x}^B \in \left\{2^{-(k+1)}\right\} \times B(w).$$

Since $\bar{x} = \lambda_A \bar{x}^A + \lambda_B \bar{x}^B$, we obtain $\bar{x} \in D^k(w)$. This means that also $C \subseteq D^k(w)$.

END

Lemma 8.2.2. The union of $D^k(w)$ over all $w \in Q$ is equal to

$$[2^{-(k+1)}, 2^{-k}] \times R^n.$$

Proof. From the construction we know that for all $w \in Q$ every element in $D^k(w)$ lies in $[2^{-(k+1)}, 2^{-k}] \times R^n$. Next, take $d = (d_0, d_1, \cdots, d_n)^\mathsf{T} \in [2^{-(k+1)}, 2^{-k}] \times R^n$ and choose w such that

$$w_i = \begin{cases} \lfloor d_i \rfloor & \text{if } \lfloor d_i \rfloor \text{ is even and } d_i - \lfloor d_i \rfloor \leq 2 - 2^{k+1} d_0 \\ & \text{or } \lfloor d_i \rfloor \text{ is odd and } d_i - \lfloor d_i \rfloor \leq 2^{k+1} d_0 - 1, \\ \lfloor d_i \rfloor + 1 & \text{otherwise,} \end{cases}$$

for $i = 1, 2, \cdots, n$. Then $\mid d_i - w_i \mid \leq 2^{k+1} d_0 - 1$ for $i \in I_o(w)$ and $\mid d_i - w_i \mid \leq 2 - 2^{k+1} d_0$ for $i \in I_e(w)$. Thus $d \in D^k(w)$. The lemma follows immediately.

$$\textbf{END}$$

Lemma 8.2.3. For $w^1, w^2 \in Q$, $D^k(w^1) \cap D^k(w^2)$ is either empty or a common face of both $D^k(w^1)$ and $D^k(w^2)$, and when $D^k(w^1) \cap D^k(w^2)$ is not empty, then it is equal to the set G defined by the convex hull of the set

$$\left(\{2^{-k}\} \times (A(w^1) \cap A(w^2))\right) \cup \left(\{2^{-(k+1)}\} \times (B(w^1) \cap B(w^2))\right).$$

Proof. Suppose that $D^k(w^1) \cap D^k(w^2)$ is not empty. Obviously,

$$G \subseteq D^k(w^1) \cap D^k(w^2).$$

Let $\bar{x} = (x_0, x)^\mathsf{T} \in D^k(w^1) \cap D^k(w^2)$. Then there exist $\lambda^i_A \geq 0$, $\lambda^i_B \geq 0$,

$$\bar{x}^i_A = (2^{-k}, x^i_A)^\mathsf{T} \in \{2^{-k}\} \times A(w^i)$$

and

$$\bar{x}^i_B = (2^{-(k+1)}, x^i_B)^\mathsf{T} \in \{2^{-(k+1)}\} \times B(w^i)$$

such that

$$\lambda^i_A + \lambda^i_B = 1 \text{ and } \bar{x} = \lambda^i_A \bar{x}^i_A + \lambda^i_B \bar{x}^i_B$$

for $i = 1, 2$. Thus $\lambda^1_A = \lambda^2_A$ and $\lambda^1_B = \lambda^2_B$. If $\lambda^1_A = 0$ or $\lambda^1_B = 0$, then $\bar{x} \in G$. If $\lambda^1_A > 0$ and $\lambda^1_B > 0$, then consider the following two cases for $i \in N$.

1. If $i \in I_o(w^1)$ and $i \in I_e(w^2)$ or $i \in I_e(w^1)$ and $i \in I_o(w^2)$, then

$$\mid w^1_i - w^2_i \mid = 1.$$

Thus $x^1_{Ai} = x^2_{Ai}$ and $x^1_{Bi} = x^2_{Bi}$.

2. If $i \in I_o(w^1)$ and $i \in I_o(w^2)$ or $i \in I_e(w^1)$ and $i \in I_e(w^2)$, then

$$| w_i^1 - w_i^2 | < 2.$$

Thus $x_{Ai}^1 = x_{Ai}^2$ and $x_{Bi}^1 = x_{Bi}^2$.

This means that $\bar{x}_A^1 = \bar{x}_A^2$ and $\bar{x}_B^1 = \bar{x}_B^2$. Therefore, $\bar{x} \in G$ and

$$D^k(w^1) \cap D^k(w^2) \subseteq G.$$

From these results, the lemma follows immediately.

END

The lemmas presented above can also be found in [95], but without proofs. The proofs are given here to make the following results more understandable.

Let $\bar{D}_1$ denote the set of faces of all simplices of the D_1-triangulation of R^n and for $w \in Q$ let $2^{-k}\bar{D}_1 \mid 2^{-(k+1)}A(w)$ be defined by

$$\left\{ \sigma \subseteq 2^{-(k+1)}A(w) \mid \sigma \in 2^{-k}\bar{D}_1 \text{ and } \dim(\sigma) = \dim(A(\dot{w})) \right\}$$

and let $2^{-(k+1)}\bar{D}_1 \mid 2^{-(k+1)}B(w)$ be defined by

$$\left\{ \sigma \subseteq 2^{-(k+1)}B(w) \mid \sigma \in 2^{-(k+1)}\bar{D}_1 \text{ and } \dim(\sigma) = \dim(B(w)) \right\}.$$

Then it is obvious that $2^{-k}\bar{D}_1 \mid 2^{-(k+1)}A(w)$ is a triangulation of $2^{-(k+1)}A(w)$ and $2^{-(k+1)}\bar{D}_1 \mid 2^{-(k+1)}B(w)$ is a triangulation of $2^{-(k+1)}B(w)$.

Let a denote the number of elements in the set $I_o(w)$ and b number of elements in the set $I_e(w)$. Let $\sigma_A \in 2^{-k}\bar{D}_1 \mid 2^{-(k+1)}A(w)$ be equal to the convex hull of $y_A^0, y_A^1, \cdots, y_A^a$ and let $\sigma_B \in 2^{-(k+1)}\bar{D}_1 \mid 2^{-(k+1)}B(w)$ be equal to the convex hull of $y_B^0, y_B^1, \cdots, y_B^b$. In addition, let σ denote the convex hull of the set $\left(\{2^{-k}\} \times \sigma_A \right) \cup \left(\{2^{-(k+1)}\} \times \sigma_B \right)$.

Lemma 8.2.4. σ is a simplex in $[2^{-(k+1)}, 2^{-k}] \times R^n$ and is equal to the convex hull of $(2^{-k}, y_A^0)^\mathsf{T}, (2^{-k}, y_A^1)^\mathsf{T}, \ldots, (2^{-k}, y_A^a)^\mathsf{T}, (2^{-(k+1)}, y_B^0)^\mathsf{T}, (2^{-(k+1)}, y_B^1)^\mathsf{T}, \cdots, (2^{-(k+1)}, y_B^b)^\mathsf{T}$.

Proof. We only need to show that σ is a $(n+1)$-simplex. Suppose that there exist $\lambda_0^A, \cdots, \lambda_a^A, \lambda_0^B, \cdots, \lambda_b^B$ not all equal to zero such that

$$\lambda_0^A + \cdots + \lambda_a^A + \lambda_0^B + \cdots + \lambda_b^B = 0$$

and

$$\sum_{i=0}^{a} \lambda_i^A (2^{-k}, y_A^i)^\top + \sum_{i=0}^{b} \lambda_i^B (2^{-(k+1)}, y_B^i)^\top = 0.$$

Then we conclude that at least one of

$$(2^{-k}, y_A^0)^\top, (2^{-k}, y_A^1)^\top, \cdots, (2^{-k}, y_A^a)^\top$$

and

$$(2^{-(k+1)}, y_B^0)^\top, (2^{-(k+1)}, y_B^1)^\top, \cdots, (2^{-(k+1)}, y_B^b)^\top$$

is affinely dependent. This contradicts the fact that both

$$(2^{-k}, y_A^0)^\top, (2^{-k}, y_A^1)^\top, \cdots, (2^{-k}, y_A^a)^\top$$

and

$$(2^{-(k+1)}, y_B^0)^\top, (2^{-(k+1)}, y_B^1)^\top, \cdots, (2^{-(k+1)}, y_B^b)^\top$$

are affinely independent.

END

Let $T(k, k+1)$ denote the collection of all such simplices σ constructed above.

Lemma 8.2.5. For $\sigma^1, \sigma^2 \in T(k, k+1)$, $\sigma^1 \cap \sigma^2$ is either empty or a common face of both σ^1 and σ^2.

Proof. Clearly, σ^1 is equal to the convex hull of the set $(\{2^{-k}\} \times \tau_A^1) \cup (\{2^{-(k+1)}\} \times \tau_B^1)$ for some τ_A^1 in $2^{-k} \bar{D}_1$ and some τ_B^1 in $2^{-(k+1)} \bar{D}_1$ and σ^2 is equal to the convex hull of the set $(\{2^{-k}\} \times \tau_A^2) \cup (\{2^{-(k+1)}\} \times \tau_B^2)$ for some τ_A^2 in $2^{-k} \bar{D}_1$ and some τ_B^2 in $2^{-(k+1)} \bar{D}_1$. From **Lemma 8.2.3**, we have that $\sigma^1 \cap \sigma^2$ is equal to the convex hull of the set $(\{2^{-k}\} \times (\tau_A^1 \cap \tau_A^2)) \cup (\{2^{-(k+1)}\} \times (\tau_B^1 \cap \tau_B^2))$. Since $\tau_A^1 \cap \tau_A^2$ is either empty or a common face of both τ_A^1 and τ_A^2 and $\tau_B^1 \cap \tau_B^2$ is either empty or a common face of both τ_B^1 and τ_B^2, the lemma follows immediately.

END

Lemma 8.2.6. The union of all simplices in $T(k, k+1)$ is equal to

$$[2^{-(k+1)}, 2^{-k}] \times R^n.$$

Proof. From **Lemma 8.2.2**, the lemma follows immediately.

END

Lemma 8.2.7. $T(k, k+1)$ triangulates $[2^{-(k+1)}, 2^{-k}] \times R^n$.

Proof. From **Lemma 8.2.4**, **Lemma 8.2.5**, and **Lemma 8.2.6**, the lemma follows immediately.

END

Theorem 8.2.8. The union of $T(k, k+1)$ over all nonnegative integers k triangulates $(0, 1] \times R^n$.

Proof. The result is clear, the proof is omitted.

END

Next we show that the set D_3 of simplices obtained in Section 2 coincides with the union of $T(k, k+1)$ over all nonnegative integers k. It is obvious that if σ is a simplex in $T(k, K+1)$ for some integer $k \geq 0$ then σ is equal to $D_3(y, \pi, s, p_1, p_2)$ for some y, π, s, p_1 and p_2.

Lemma 8.2.9. Let $y^{-1}, y^0, \cdots, y^n$ be obtained from **Definition 8.1.1** for some given y, π, s, p_1 and p_2 and let σ denote the convex hull of $y^{-1}, y^0, \cdots, y^n$. Then σ is a simplex in $T(k, k+1)$ with k such that $y_0 = 2^{-(k+1)}$.

Proof. From the definition of $y^{-1}, y^0, \cdots, y^n$, we can rewrite them into

$$y^{-1} = (y_0, \tilde{y}^{-1})^\mathsf{T}, \quad \cdots, \quad y^{j-1} = (y_0, \tilde{y}^{j-1})^\mathsf{T}$$

and

$$y^j = (2y_0, \tilde{y}^j)^\mathsf{T}, \quad \cdots, \quad y^n = (2y_0, \tilde{y}^n)^\mathsf{T}.$$

Let $\tilde{\tau}_A$ be equal to the convex hull of $\tilde{y}^j, \cdots, \tilde{y}^n$ and let $\tilde{\tau}_B$ be equal to the convex hull of $\tilde{y}^{-1}, \cdots, \tilde{y}^{j-1}$. Furthermore, let

$$w_{\pi(i)} = (y_{\pi(i)} + y_0 s_{\pi(i)})/y_0, \quad i = 0, 1, \cdots, j-1$$

and

$$w_{\pi(i)} = y_{\pi(i)}/y_0, \quad i = j+1, \cdots, n.$$

Then it is obvious that

$$\tilde{\tau}_A \in 2y_0 \bar{D}_1 \mid y_0 A(w) \text{ and } \tilde{\tau}_B \in y_0 \bar{D}_1 \mid y_0 B(w).$$

Since σ is equal to the convex hull of $(\{2y_0\} \times \tilde{\tau}_A) \cup (\{y_0\} \times \tilde{\tau}_B)$, the lemma follows immediately.

END

Theorem 8.2.10. The set D_3 is equal to the union of $T(k, k+1)$ over all nonnegative integers k.

Proof. From the definition of the D_1-triangulation, the construction of $T(k, k+1)$, **Lemma 8.2.9**, and the definition of D_3, we obtain this conclusion immediately.

END

Theorem 8.2.10 implies that D_3 is a triangulation of $(0, 1] \times R^n$. we call it the D_3-triangulation of $(0, 1] \times R^n$.

8.3   Pivot Rules of the D_3-Triangulation

A simplicial homotopy algorithm, based on the D_3-triangulation of continuous refinement of grid sizes of $(0, 1] \times R^n$, can be implemented according to the following pivot rules. Let a simplex of the D_3-triangulation, $\sigma = D_3(y, \pi, s, p_1, p_2)$, be given with vertices $y^{-1}, y^0, \cdots, y^n$. We wish to obtain the simplex of the D_3-triangulation, $\bar{\sigma} = D_3(\bar{y}, \bar{\pi}, \bar{s}, \bar{p}_1, \bar{p}_2)$, such that all vertices of σ are also vertices of $\bar{\sigma}$ except y^i. **Table 8.3.1** describes how $\bar{y}, \bar{\pi}, \bar{s}, \bar{p}_1$ and $\bar{p}_2$ depend on y, π, s, p_1, p_2 and i. From this table, it is easy to obtain all vertices of $\bar{\sigma}$, in particular the vertex of $\bar{\sigma}$ opposite to the facet shared with σ. In this table,

$$i^* = \begin{cases} j-1 & \text{if } i = j-2, \\ j-2 & \text{if } i = j-1 \end{cases}$$

and

$$i^{**} = \begin{cases} n & \text{if } i = n-1, \\ n-1 & \text{if } i = n. \end{cases}$$

Table 8.3.1(1). The Pivot Rules of the D_3-Triangulation

i	j	p_1	p_2	condition	$\bar{y}$	$\bar{s}$	$\bar{\pi}$	$\bar{p}_1$	$\bar{p}_2$	$\bar{j}$
-1	0				$y - y_0 s$	s	$(\pi(1),\dots,\pi(n),\pi(0))$	$p_2 - 1$	0	0
	1				$y + 2y_0\, s_{\pi(0)}u^{\pi(0)}$	$s - 2s_{\pi(0)}u^{\pi(0)}$	π	p_1	p_2	n
	$j>1$	-1			y	s	π	$p_1 + 1$	p_2	j
		0			y	s	π	$p_1 - 1$	p_2	j
		$p_1 \ge 1$	$p_2 \ge 1$		y	$s - 2s_{\pi(0)}u^{\pi(0)}$	$(\pi(1),\dots,\pi(n),\pi(0))$	$p_1 - 1$	p_2	$j-1$
			$h=1$		y	$s - 2s_{\pi(0)}u^{\pi(0)}$	$(\pi(1),\dots,\pi(j),\pi(0),\pi(j+1),\dots,\pi(n))$	$p_1 - 1$	$p_2 + 1$	$j-1$
			$h>1$	$h=0$	y	s	$(\pi(1),\dots,\pi(n),\pi(0))$	$p_1 - 1$	$p_2 + 1$	$j-1$
				$h \ge 1$, $s_{\pi(0)} = t_{\pi(0)}$	y	s	$(\pi(1),\dots,\pi(n),\pi(0))$	$p_1 - 1$	$p_2 + 1$	$j-1$
				$h \ge 1$, $s_{\pi(0)} = -t_{\pi(0)}$	y	$s - 2s_{\pi(0)}u^{\pi(0)}$	$(\pi(1),\dots,\pi(j),\pi(0),\pi(j+1),\dots,\pi(n))$	$p_1 - 1$	$p_2 + 1$	$j-1$
$0 \le i \le j-1$		-1	0	$h=0$	y	$s - 2s_{\pi(i)}u^{\pi(i)}$	$(\pi(0),\dots,\pi(i-1),\pi(i+1),\dots,\pi(n),\pi(i))$	p_1	p_2	$j-1$
			$p_2 \ge 1$		y	$s - 2s_{\pi(i)}u^{\pi(i)}$	$(\pi(0),\dots,\pi(i-1),\pi(i+1),\dots,\pi(j),\pi(i),\pi(j+1),\dots,\pi(n))$	p_1	$p_2 + 1$	$j-1$
			$h=1$		y	s	$(\pi(0),\dots,\pi(i-1),\pi(i+1),\dots,\pi(n),\pi(i))$	p_1	p_2	$j-1$
			$h>1$	$h \ge 1$, $s_{\pi(i)} = t_{\pi(i)}$	y	s	$(\pi(0),\dots,\pi(i-1),\pi(i+1),\dots,\pi(n),\pi(i))$	p_1	$p_2 + 1$	$j-1$
				$h \ge 1$, $s_{\pi(i)} = -t_{\pi(i)}$	y	$s - 2s_{\pi(i)}u^{\pi(i)}$	$(\pi(0),\dots,\pi(i-1),\pi(i+1),\dots,\pi(j),\pi(i),\pi(j+1),\dots,\pi(n))$	p_1	$p_2 + 1$	$j-1$

Table 8.3.1(2). The Pivot Rules of the D_3-Triangulation

i / j ($0\le i\le j-1$)	h	range	p_1	p_2	$\bar{y}$	$\bar{s}$	$\bar{\pi}$	$\bar{p}_1$	$\bar{p}_2$	$\bar{j}$
i			$i<p_1-1$		y	s	$(\pi(0),\ldots,\pi(i+1),\pi(i),\ldots,\pi(n))$	p_1	p_2	j
			$i=p_1-1$			s	π	p_1-1	p_2	j
			$i\ge p_1$, $0\vee p_1 < j-2$		y	s	$(\pi(0),\ldots,\pi(p_1-1),\pi(i),\pi(p_1),\ldots,\pi(i-1),\pi(i+1),\ldots,\pi(n))$	p_1+1	p_2	j
			$i\ge p_1$, $0\vee p_1 = j-2$		$y+2y_0\,s_{\pi(i^*)}u^{\pi(i^*)}$	$s-2s_{\pi(i^*)}u^{\pi(i^*)}$				
j	$h=0$			0	y	s	π	p_1	p_2+1	j
		$j<n-1$	-1		y	s	$(\pi(0),\ldots,\pi(j+1),\pi(j),\ldots,\pi(n))$	p_1	p_2	j
			$p_1\ge0$		y	s	$(\pi(j+1),\pi(0),\ldots,\pi(j),\pi(j+2),\ldots,\pi(n))$	p_1+1	p_2	$j+1$
		$n-1$	-1	1	y	$s-2s_{\pi(j+1)}u^{\pi(j+1)}$	π	p_1	p_2-1	$j+1$
	$h\ge1$	$j<n$	-1	$p_2\ge2$	y	$s-2s_{\pi(j+1)}u^{\pi(j+1)}$	$(\pi(0),\ldots,\pi(j+1),\pi(j),\ldots,\pi(n))$	p_1	p_2-1	$j+1$
			$p_1\ge0$		y	$s-2s_{\pi(j+1)}u^{\pi(j+1)}$	$(\pi(j+1),\pi(0),\ldots,\pi(j),\pi(j+2),\ldots,\pi(n))$	p_1+1	p_2-1	$j+1$
			-1		y	s	$(\pi(0),\ldots,\pi(j+1),\pi(j),\ldots,\pi(n))$	p_1	p_2-1	$j+1$
			$p_1\ge0$		y	$s-2s_{\pi(j+1)}u^{\pi(j+1)}$	$(\pi(j+1),\pi(0),\ldots,\pi(j),\pi(j+2),\ldots,\pi(n))$	p_1+1	p_2-1	$j+1$
		n			$y+y_0 s/2$	s	$(\pi(n),\pi(0),\ldots,\pi(n-1))$	-1	p_1+1	0

Table 8.3.1(3). The Pivot Rules of the D_3-Triangulation

i	j	h	p_1	p_2	$\bar y$	$\bar s$	$\bar\pi$	$\bar p_1$	$\bar p_2$	$\bar j$
$j < i \leq n$		$h = 0$	-1	0	y	$s - 2s_{\pi(i)}u^{\pi(i)}$	$(\pi(0),\ldots,\pi(j-1),\pi(i),\ \pi(j),\ldots,\pi(i-1),\ \pi(i+1),\ldots,\pi(n))$	p_1	p_2	$j+1$
			$p_1 \geq 0$		y	$s - 2s_{\pi(i)}u^{\pi(i)}$	$(\pi(i),\pi(0),\ldots,\pi(i-1),\ \pi(i+1),\ldots,\pi(n))$	p_1+1	p_2	$j+1$
				$i < p_2 + j - 1$	y	s	$(\pi(0),\ldots,\pi(i+1),\ \pi(i),\ldots,\pi(n))$	p_1	p_2	j
				$i = p_2 + j - 1$	y	s	π	p_1	$p_2 - 1$	j
				$i \geq p_2 + j$, $1 \leq p_2 < n - j - 1$	y	s	$(\pi(0),\ldots,\pi(j+p_2-1),\ \pi(i),\pi(j+p_2),\ldots,\pi(i-1),\ \pi(i+1),\ldots,\pi(n))$	p_1	$p_2 + 1$	j
			-1	$i \geq n - 1$, $1 \leq p_2 = n - j$	y	s	$(\pi(0),\ldots,\pi(j-1),\ \pi(i^{**}),\pi(j),\ldots,\pi(i^{**}-1),\ \pi(i^{**}+1),\ldots,\pi(n))$	p_1	p_2	$j+1$
			$p_1 \geq 0$	-1	y	s	$(\pi(i^{**}),\pi(0),\ldots,\pi(i^{**}-1),\ \pi(i^{**}+1),\ldots,\pi(n))$	p_1+1	p_2	$j+1$
$j < i < n$		$h \geq 1$			y	s	$(\pi(0),\ldots,\pi(i+1),\ \pi(i),\ldots,\pi(n))$	p_1	p_2	j
n	$j < n$		-1		y	s	$(\pi(0),\ldots,\pi(j-1),\ \pi(n),\pi(j),\ldots,\pi(n-1))$	p_1	$p_2 - 1$	$j+1$
			$p_1 \geq 0$	-1	y	s	$(\pi(n),\pi(0),\ldots,\pi(n-1))$	p_1+1	$p_2 - 1$	$j+1$

8.4   Comparison of Several Triangulations for Simplicial Homotopy Algorithms

In this section we first calculate the number of simplices of different continuous grid refinement triangulations in the set $[1/2, 1] \times U^n$, where U^n is the n-dimensional unit cube. For the definitions of the K_3-triangulation and of the J_3-triangulation, see Chapter 3.

Theorem 8.4.1. The number of simplices of the K_3-triangulation and of the J_3-triangulation in the set $[1/2, 1] \times U^n$ is equal to

$$p_n = (2^{n+1} - 1)n!.$$

Proof. See [177].

END

Theorem 8.4.2. The number of simplices of the D_3-triangulation in the set $[1/2, 1] \times U^n$ is equal to

$$q_n = \sum_{m=0}^{n} ((2^m - 1)C_n^m d_m(n - m)! + C_n^m d_m d_{n-m}),$$

where

$$d_j = j + j(j - 1) + \cdots + j(j - 1) \cdots 4 \cdot 3 + 2$$

for $j \geq 2$, $d_0 = d_1 = 1$, and $C_n^m = \frac{n!}{m!(n-m)!}$.

Proof. Let $\bar{Q}$ denote the set $\{w \in R^n \mid w_i \in \{0, 1, 2\}$ for $i = 1, 2, \cdots, n\}$. Take $w \in \bar{Q}$. Let $\bar{A}(w)$ denote the set

$$\left\{ x \in R^n \;\middle|\; \begin{array}{l} w_i - 1 \leq x_i \leq w_i + 1 \text{ for } i \in I_o(w) \text{ and} \\ x_i = w_i \text{ for } i \in I_e(w) \end{array} \right\}$$

and let $\bar{B}(w)$ denote the set

$$\left\{ x \in R^n \;\middle|\; \begin{array}{l} x_i = w_i \text{ for } i \in I_o(w), \\ w_i \leq x_i \leq w_i + 1 \text{ for } i \in I_e(w) \text{ and } w_i = 0, \\ w_i - 1 \leq x_i \leq w_i \text{ for } i \in I_e(w) \text{ and } w_i = 2 \end{array} \right\}.$$

Furthermore, let $1/2\bar{D}(w)$ denote the convex hull of the set

$$(\{1\} \times 1/2\bar{A}(w)) \cup (\{1/2\} \times 1/2\bar{B}(w)).$$

Then it is obvious that

$$[1/2, 1] \times U^n = \cup_{w \in \bar{Q}} 1/2\bar{D}(w).$$

Let m denote the number of elements in $I_e(w)$. Then there are $2^m C_n^m$ elements in $\bar{Q}$ such that m components of each of them are even. Thus the number of simplices of the D_3-triangulation in the set

$$\cup_{w \in \bar{Q}, |I_e(w)|=m} 1/2\bar{D}(w)$$

is equal to

$$(2^m - 1)C_n^m d_m(n - m)! + C_n^m d_m d_{n-m}.$$

Since

$$\cup_{m=0}^n (\cup_{w \in \bar{Q}, |I_e(w)|=m} 1/2\bar{D}(w)) = [1/2, 1] \times U^n,$$

the theorem follows immediately.

END

Theorem 8.4.3 When $n \geq 3$, $q_n < p_n$. And as n goes to infinity, q_n/p_n converges to $e - 2$.

Proof. The conclusion is obvious, the proof is omitted.

END

From **Theorem 8.4.3**, we conclude that the number of simplices of the D_3-triangulation is smaller than for the other triangulations with a fixed refinement factor of two that underlie simplicial homotopy algorithms. Furthermore, since the D_3-triangulation subdivides each level $\{2^{-k}\} \times R^n$, for $k = 0, 1, \cdots$, into simplices according to the D_1-triangulation of R^n with grid size 2^{-k}, the average directional density of the D_3-triangulation is on each level less than the one of the K_3-triangulation or of the J_3-triangulation. The latter two triangulations subdivide each level $\{2^{-k}\} \times R^n$ into simplices, for $k = 0, 1, \cdots$, according to the K_1-triangulation of R^n and the J_1-triangulation of R^n with grid size 2^{-k}, respectively.

Chapter 9

The D_2-Triangulation for Simplicial Homotopy Algorithms

Triangulations of continuous refinement of grid sizes with a fixed factor of two for $(0,1] \times R^n$ are simple, but they can not reach quadratic convergence when being used to compute zero points of smooth nonlinear mappings. In this chapter we develop a new triangulation of continuous refinement of grid sizes with arbitrary factors for $(0,1] \times R^n$, based on the D_1-triangulation. This simplicial subdivision is called the D_2-triangulation. We show that the D_2-triangulation is superior to the K_2-triangulation and the J_2-triangulation proposed by Kojima and Yamamoto in [95] according to measures of efficiency. However, the D_2-triangulation does not induce the D_3-triangulation as its special case. Therefore, we also give a new triangulation of continuous refinement of grid sizes with even factors for $(0,1] \times R^n$ such that the D_3-triangulation is derived as its special case. This simplicial subdivision is called the D_2^*-triangulation. According to measures of efficiency we also show that the D_2^*-triangulation is superior to the K_2^*-triangulation and the J_2^*-triangulation proposed by Kojima and Yamamoto in [95]. The latter two triangulations are such that the K_3-triangulation and the J_3-triangulation are induced as their special case, respectively. This chapter is organized as follows. Section 1 gives the constructive procedure of the D_2-triangulation. The algebraic definition of the D_2-triangulation is introduced in Section 2. The pivot rules of the D_2-triangulation is described in Section 3. Section 4 gives the definition of D_2^*-triangulation. We describe the pivot rules of the D_2^*-triangulation in Section 5. Finally, several triangulations with continuous grid refinement for $(0, 1] \times R^n$ are compared in Section 6. This chapter is based on Dang's [15] and [16].

9.1   Construction of the D_2-Triangulation

Let n be a positive integer and let N denote the index set $\{1, 2, \cdots, n\}$. Next, let Q denote the set of vectors in R^n whose components are integers. Take an arbitrary element $w \in Q$. Then define

$$I_o(w) = \{i \in N \mid w_i \text{ is odd}\} \text{ and } I_e(w) = \{j \in N \mid w_j \text{ is even}\}.$$

Furthermore, let $A(w)$ denote the set

$$\{x \in R^n \mid w_i - 1 \le x_i \le w_i + 1 \text{ for } i \in I_o(w) \text{ and } x_i = w_i \text{ for } i \in I_e(w)\}$$

and let $B(w)$ denote the set

$$\{x \in R^n \mid x_i = w_i \text{ for } i \in I_o(w) \text{ and } w_i - 1 \le x_i \le w_i + 1 \text{ for } i \in I_e(w)\}.$$

Let k be a nonnegative integer. Then let $D^k(w)$ denote the convex hull of the set

$$(\{2^{-k}\} \times A(w)) \cup (\{2^{-(k+1)}\} \times B(w)).$$

As were shown in Chapter 8,

$$D^k(w) = \left\{d \in [2^{-(k+1)}, 2^{-k}] \times R^n \left| \begin{array}{l} |d_i - w_i| \le 2^{k+1}d_0 - 1 \text{ for } i \in I_o(w) \\ |d_i - w_i| \le 2 - 2^{k+1}d_0 \text{ for } i \in I_e(w) \end{array} \right.\right\}$$

and

$$\cup_{w \in Q} D^k(w) = [2^{-(k+1)}, 2^{-k}] \times R^n.$$

For $w^1, w^2 \in Q$, $D^k(w^1) \cap D^k(w^2)$ is either empty or a common face of both $D^k(w^1)$ and $D^k(w^2)$, and when $D^k(w^1) \cap D^k(w^2)$ is not empty, it is equal to the convex hull of the set

$$(\{2^{-k}\} \times (A(w^1) \cap A(w^2))) \cup (\{2^{-(k+1)}\} \times (B(w^1) \cap B(w^2))).$$

Take G to be one out of the K_1-triangulation, the J_1-triangulation or the D_1-triangulation of R^n. Let $\bar{G}$ denote the set of faces of all simplices in G. Then take $\alpha_0 \in (0, 1]$ and $\beta_i \in \{1/j \mid j = 1, 2, \cdots\}$ for $i = 0, 1, \cdots$. Take α_{j+1} such that $\alpha_{j+1} = \alpha_j \beta_j$, for $j = 0, 1, \cdots$.

Let $\alpha_k \bar{G} \mid \alpha_k A(w)$ be defined by the set

$$\left\{ \sigma \subseteq \alpha_k A(w) \mid \sigma \in \alpha_k \bar{G} \text{ and } \dim(\sigma) = \dim(A(w)) \right\}$$

and let $\alpha_{k+1} \bar{G} \mid \alpha_k B(w)$ be defined by the set

$$\left\{ \sigma \subseteq \alpha_k B(w) \mid \sigma \in \alpha_{k+1} \bar{G} \text{ and } \dim(\sigma) = \dim(B(w)) \right\}.$$

For the D_1-triangulation, the K_1-triangulation, and the J_1-triangulation, it is obvious that $\alpha_k \bar{G} \mid \alpha_k A(w)$ is a triangulation of $\alpha_k A(w)$ and $\alpha_{k+1} \bar{G} \mid \alpha_k B(w)$ is a triangulation of $\alpha_k B(w)$.

Let a denote the number of elements in the set $I_o(w)$ and b the number of elements in the set $I_e(w)$. Next, let $\sigma_A \in \alpha_k \bar{G} \mid \alpha_k A(w)$ with vertices $y_A^0, y_A^1, \cdots, y_A^a$ and let $\sigma_B \in \alpha_{k+1} \bar{G} \mid \alpha_k B(w)$ with vertices $y_B^0, y_B^1, \cdots, y_B^b$. Furthermore, let σ denote the convex hull of the set

$$\left(\left\{ 2^{-k} \right\} \times \sigma_A \right) \cup \left(\left\{ 2^{-(k+1)} \right\} \times \sigma_B \right).$$

Then it can easily be obtained that σ is a simplex and is equal to the convex hull of $(2^{-k}, y_A^0)^\top$, $(2^{-k}, y_A^1)^\top$, $\cdots$, $(2^{-k}, y_A^a)^\top$, $(2^{-(k+1)}, y_B^0)^\top$, $(2^{-(k+1)}, y_B^1)^\top$, $\cdots$, $(2^{-(k+1)}, y_B^b)^\top$.

Let $T(k, k+1)$ denote the collection of all such simplices σ. Clearly, we have that for $\sigma^1, \sigma^2 \in T(k, k+1)$, $\sigma^1 \cap \sigma^2$ is either empty or a common face of both σ^1 and σ^2. Moreover, $\cup_{\sigma \in T(k, k+1)} \sigma = [2^{-(k+1)}, 2^{-k}] \times R^n$. Hence $T(k, k+1)$ is a triangulation of $[2^{-(k+1)}, 2^{-k}] \times R^n$.

Theorem 9.1.1. $\cup_{k=0}^\infty T(k, k+1)$ is a triangulation of $(0, 1] \times R^n$.

Proof. From the choice of α_j and β_j for $j = 0, 1, \cdots$, we obtain this conclusion.

END

We call the triangulation constructed in **Theorem 9.1.1** the G_2-triangulation. In this way we obtain the K_2-triangulation, the J_2-triangulation, and the D_2-triangulation of $(0, 1] \times R^n$. In case of the

D_2-triangulation, each level $\{2^{-k}\} \times R^n$ is triangulated according to the D_1-triangulation with grid size α_k for $k = 0, 1, \cdots$. Similarly for the K_2-triangulation and the J_2-triangulation.

9.2 Description of the D_2-Triangulation

Let N_0 denote the index set $\{0, 1, \cdots, n\}$ and let u^i denote the i-th unit vector in R^{n+1} for $i = 0, 1, \cdots, n$.

Take a permutation $\pi = (\pi(0), \pi(1), \cdots, \pi(n))$ of the elements of N_0. Let q denote the integer such that $\pi(q) = 0$. Take a vector y in $(0, 1] \times R^n$ such that for some nonnegative integer k, $y_0 = 2^{-(k+1)}$ and $y_{\pi(i)}/\alpha_{k+1}$ is an integer for $i = 0, \cdots, q-1$ and $y_{\pi(i)}/\alpha_k$ is an integer for $i = q + 1, \cdots, n$. Then define

$$w_{\pi(i)} = \begin{cases} \lfloor y_{\pi(i)}/\alpha_k \rfloor + 1 & \text{if } \lfloor y_{\pi(i)}/\alpha_k \rfloor \text{ is odd,} \\ \lfloor y_{\pi(i)}/\alpha_k \rfloor & \text{otherwise,} \end{cases}$$

for $i = 0, 1, \cdots, q - 1$ and

$$w_{\pi(i)} = \begin{cases} y_{\pi(i)}/\alpha_k + 1 & \text{if } y_{\pi(i)}/\alpha_k \text{ is even,} \\ y_{\pi(i)}/\alpha_k & \text{otherwise,} \end{cases}$$

for $i = q + 1, \cdots, n$.

Definition 9.2.1. Let y and π be as given above. Then y^{-1}, y^0, $\cdots$, y^n are given as follows:

$$\begin{aligned} y^{-1} &= \sum_{j=0}^{q} y_{\pi(j)} u^{\pi(j)} + \alpha_k \sum_{j=q+1}^{n} w_{\pi(j)} u^{\pi(j)}, \\ y^i &= y^{i-1} + \alpha_{k+1} u^{\pi(i)}, \quad i = 0, 1, \cdots, q - 1, \\ y^q &= \alpha_k \sum_{j=0}^{q-1} w_{\pi(j)} u^{\pi(j)} + \sum_{j=q+1}^{n} y_{\pi(j)} u^{\pi(j)} + 2 y_0 u^0, \\ y^i &= y^{i-1} + \alpha_k u^{\pi(i)}, \quad i = q + 1, \cdots, n. \end{aligned}$$

Let y^{-1}, y^0, $\cdots$, y^n be obtained from the above definition. Then it is obvious that they are affinely independent. Thus their convex hull is a simplex. Let us denote this simplex by $K_2(y, \pi)$. Then the K_2-triangulation is the set of all such simplices $K_2(y, \pi)$. Following the conclusions in the previous section, we have that this triangulation is a grid refinement simplicial subdivision of $(0,1] \times R^n$ such that the factor of refinement can be chosen arbitrarily between any two levels.

Take a permutation $\pi = (\pi(0), \pi(1), \cdots, \pi(n))$ of the elements of N_0. Let q denote the integer such that $\pi(q) = 0$. Take a vector y in $(0,1] \times R^n$ such that for some nonnegative integer k, $y_0 = 2^{-(k+1)}$ and either $y_{\pi(i)}/\alpha_k$ is even for $i = q+1, \cdots, n$ and $y_{\pi(i)}/\alpha_{k+1}$ is even for $i = 0, \cdots, q-1$ or $y_{\pi(i)}/\alpha_k$ is odd for $i = q+1, \cdots, n$ and if $1/\beta_k$ is even, $y_{\pi(i)}/\alpha_{k+1}$ is even for $i = 0, \cdots, q-1$ and if $1/\beta_k$ is odd, $y_{\pi(i)}/\alpha_{k+1}$ is odd for $i = 0, \cdots, q-1$. Take s to be a sign vector. If $y_{\pi(j)}/\alpha_k$ is odd for $j = q+1, \cdots, n$, define

$$
w_{\pi(i)} = \begin{cases} \lfloor y_{\pi(i)}/\alpha_k \rfloor + 1 & \text{if } \lfloor y_{\pi(i)}/\alpha_k \rfloor \text{ is odd and} \\ & \text{either } y_{\pi(i)}/\alpha_k \neq \lfloor y_{\pi(i)}/\alpha_k \rfloor \\ & \text{or both } \lfloor y_{\pi(i)}/\alpha_k \rfloor = y_{\pi(i)}/\alpha_k \text{ and } s_{\pi(i)} = 1, \\ \lfloor y_{\pi(i)}/\alpha_k \rfloor & \text{if } \lfloor y_{\pi(i)}/\alpha_k \rfloor \text{ is even,} \\ \lfloor y_{\pi(i)}/\alpha_k \rfloor - 1 & \text{otherwise,} \end{cases}
$$

for $i = 0, 1, \cdots, q-1$ and if $y_{\pi(j)}/\alpha_k$ is even for $j = q+1, \cdots, n$, define

$$
w_{\pi(i)} = \begin{cases} \lfloor y_{\pi(i)}/\alpha_k \rfloor + 1 & \text{if } \lfloor y_{\pi(i)}/\alpha_k \rfloor \text{ is even and} \\ & \text{either } y_{\pi(i)}/\alpha_k \neq \lfloor y_{\pi(i)}/\alpha_k \rfloor \\ & \text{or both } \lfloor y_{\pi(i)}/\alpha_k \rfloor = y_{\pi(i)}/\alpha_k \text{ and } s_{\pi(i)} = 1, \\ \lfloor y_{\pi(i)}/\alpha_k \rfloor & \text{if } \lfloor y_{\pi(i)}/\alpha_k \rfloor \text{ is odd,} \\ \lfloor y_{\pi(i)}/\alpha_k \rfloor - 1 & \text{otherwise,} \end{cases}
$$

for $i = 0, 1, \cdots, q-1$.

Definition 9.2.2. Let y, π and s be as given above. Then y^{-1}, y^0, $\cdots$, y^n are given as follows:

$$
\begin{aligned}
y^{-1} &= y, \\
y^i &= y^{i-1} + \alpha_{k+1} s_{\pi(i)} u^{\pi(i)}, \quad i = 0, 1, \cdots, q-1, \\
y^q &= \alpha_k \sum_{j=0}^{q-1} w_{\pi(j)} u^{\pi(j)} + \sum_{j=q+1}^{n} (y_{\pi(j)} - \alpha_k s_{\pi(j)}) u^{\pi(j)} + 2y_0 u^0, \\
y^i &= y^{i-1} + \alpha_k s_{\pi(i)} u^{\pi(i)}, \quad i = q+1, \cdots, n.
\end{aligned}
$$

Let y^{-1}, y^0, $\cdots$, y^n be obtained from the above definition. Then it is obvious that they are affinely independent. Thus their convex hull is a simplex. Let us denote this simplex by $J_2(y, \pi, s)$. Then the J_2-triangulation is the set of all such simplices $J_2(y, \pi, s)$. Following the conclusions in the previous section, we have that this triangulation is a grid refinement simplicial subdivision of $(0,1] \times R^n$ such that the factor of refinement can be chosen arbitrarily between any two levels.

Take a permutation $\pi = (\pi(0), \pi(1), \cdots, \pi(n))$ of the elements of N_0. Let q denote the integer such that $\pi(q) = 0$. Take a vector y in $(0,1] \times R^n$ such that for some nonnegative integer k, $y_0 = 2^{-(k+1)}$ and either $y_{\pi(i)}/\alpha_k$ is even for $i = q+1, \cdots, n$ and $y_{\pi(i)}/\alpha_{k+1}$ is even for $i = 0, \cdots, q-1$ or $y_{\pi(i)}/\alpha_k$ is odd for $i = q+1, \cdots, n$ and if $1/\beta_k$ is even, $y_{\pi(i)}/\alpha_{k+1}$ is even for $i = 0, \cdots, q-1$ and if $1/\beta_k$ is odd, $y_{\pi(i)}/\alpha_{k+1}$ is odd for $i = 0, \cdots, q-1$. Take s to be a sign vector. If $y_{\pi(j)}/\alpha_k$ is odd for $j = q+1, \cdots, n$, define

$$
w_{\pi(i)} = \begin{cases}
\lfloor y_{\pi(i)}/\alpha_k \rfloor + 1 & \text{if } \lfloor y_{\pi(i)}/\alpha_k \rfloor \text{ is odd and} \\
& \text{either } y_{\pi(i)}/\alpha_k \neq \lfloor y_{\pi(i)}/\alpha_k \rfloor \\
& \text{or both } \lfloor y_{\pi(i)}/\alpha_k \rfloor = y_{\pi(i)}/\alpha_k \text{ and } s_{\pi(i)} = 1, \\
\lfloor y_{\pi(i)}/\alpha_k \rfloor & \text{if } \lfloor y_{\pi(i)}/\alpha_k \rfloor \text{ is even,} \\
\lfloor y_{\pi(i)}/\alpha_k \rfloor - 1 & \text{otherwise,}
\end{cases}
$$

for $i = 0, 1, \cdots, q-1$ and if $y_{\pi(j)}/\alpha_k$ is even for $j = q+1, \cdots, n$, define

$$
w_{\pi(i)} = \begin{cases}
\lfloor y_{\pi(i)}/\alpha_k \rfloor + 1 & \text{if } \lfloor y_{\pi(i)}/\alpha_k \rfloor \text{ is even and} \\
& \text{either } y_{\pi(i)}/\alpha_k \neq \lfloor y_{\pi(i)}/\alpha_k \rfloor \\
& \text{or both } \lfloor y_{\pi(i)}/\alpha_k \rfloor = y_{\pi(i)}/\alpha_k \text{ and } s_{\pi(i)} = 1, \\
\lfloor y_{\pi(i)}/\alpha_k \rfloor & \text{if } \lfloor y_{\pi(i)}/\alpha_k \rfloor \text{ is odd,} \\
\lfloor y_{\pi(i)}/\alpha_k \rfloor - 1 & \text{otherwise,}
\end{cases}
$$

for $i = 0, 1, \cdots, q-1$.

Take two integers p_1 and p_2 such that $-1 \leq p_1 \leq q-2$ and $0 \leq p_2 \leq n-q-1$.

Definition 9.2.3. Let y, π, s, p_1 and p_2 be as given above. Then y^{-1}, y^0, $\cdots$, y^n are given as follows.

When $p_1 = -1$, $y^{-1} = y$,

$$y^i = y + \alpha_{k+1} s_{\pi(i)} u^{\pi(i)}, \quad i = 0, 1, \cdots, q-1,$$

and when $p_1 \geq 0$,

$$y^{-1} = y + \alpha_{k+1} \sum_{j=0}^{q-1} s_{\pi(j)} u^{\pi(j)},$$
$$y^i = y^{i-1} - \alpha_{k+1} s_{\pi(i)} u^{\pi(i)}, \quad i = 0, 1, \cdots, p_1 - 1,$$
$$y^i = y + \alpha_{k+1} s_{\pi(i)} u^{\pi(i)}, \quad i = p_1, \cdots, q - 1.$$

When $p_2 = 0$,

$$y^q = \alpha_k \sum_{j=0}^{q-1} w_{\pi(j)} u^{\pi(j)} + \sum_{j=q+1}^{n} (y_{\pi(j)} - \alpha_k s_{\pi(j)}) u^{\pi(j)} + 2 y_0 u^0,$$
$$y^i = y^q + \alpha_k s_{\pi(i)} u^{\pi(i)}, \quad i = q+1, \cdots, n,$$

and when $p_2 \geq 1$,

$$y^q = \alpha_k \sum_{j=0}^{q-1} w_{\pi(j)} u^{\pi(j)} + \sum_{j=q+1}^{n} y_{\pi(j)} u^{\pi(j)} + 2 y_0 u^0,$$
$$y^i = y^{i-1} - \alpha_k s_{\pi(i)} u^{\pi(i)}, \quad i = q+1, \cdots, q + p_2 - 1,$$
$$y^i = y^* + \alpha_k s_{\pi(i)} u^{\pi(i)}, \quad i = q + p_2, \cdots, n,$$

where

$$y^* = \alpha_k \sum_{j=0}^{q-1} w_{\pi(j)} u^{\pi(j)} + \sum_{j=q+1}^{n} (y_{\pi(j)} - \alpha_k s_{\pi(j)}) u^{\pi(j)} + 2 y_0 u^0.$$

Let y^{-1}, y^0, $\cdots$, y^n be obtained from the above definition. Then it is obvious that they are affinely independent. Thus their convex hull is a simplex. Let us denote this simplex by $D_2(y, \pi, s, p_1, p_2)$. Then the D_2-triangulation is the set of all such simplices $D_2(y, \pi, s, p_1, p_2)$. Following the conclusions in the previous section, we have that this triangulation is a grid refinement simplicial subdivision of $(0, 1] \times R^n$ such that the factor of refinement can be chosen arbitrarily between any two levels.

9.3 Pivot Rules of the D_2-Triangulation

Let $f : R^n \to R^n$ be a continuous function. Suppose that we want to compute a zero point of f, i.e., a vector $x^* \in R^n$ such that $f(x^*) = 0$. Let v be an arbitrary point in R^n. Then the function $g : (0, 1] \times R^n \to R^n$ is defined by $g(t, x) = (1 - t) f(x) + t(x - v)$. Let $(0, 1] \times R^n$ be triangulated according to one of the triangulations defined above, denoted by the G_2-triangulation. Next, let H be the piecewise linear approximation of g with respect to the G_2-triangulation. More precisely, let

$x = \sum_{i=-1}^{n} \lambda_i y^i$ be a vector in some simplex of the G_2-triangulation with vertices y^{-1}, y^0, $\cdots$, y^n, where $\lambda_i \geq 0$ for all i and $\sum_{i=-1}^{n} \lambda_i = 1$. Then $H(x)$ is defined by

$$H(x) = \sum_{i=-1}^{n} \lambda_i g(y^i).$$

Clearly $H(1, v) = 0$ and $H(1, w) \neq 0$ for $w \neq v$. Now the simplicial homotopy algorithm follows the piecewise linear path, P, of zero points of H originating at $(1, v)$. The path P is linear on each simplex σ of the G_2-triangulation it passes. Such a linear piece can be generated by making a linear programming step in the system of linear equations

$$\sum_{i=-1}^{n} \lambda_i \begin{pmatrix} g(y^i) \\ 1 \end{pmatrix} = \begin{pmatrix} 0 \\ 1 \end{pmatrix}.$$

If by implementing a linear programming step some λ_i becomes zero, then the vertex y^i of σ is replaced by a vertex of the unique simplex of the G_2-triangulation, say $\bar{\sigma}$, adjacent to σ and sharing with σ the facet opposite to y^i. This vertex of $\bar{\sigma}$ lies opposite to that facet.

Let a simplex of the K_2-triangulation, $\sigma = K_2(y, \pi)$, be given with vertices y^{-1}, y^0, $\cdots$, y^n. We wish to obtain the simplex of the K_2-triangulation, $\bar{\sigma} = K_2(\bar{y}, \bar{\pi})$, such that all vertices of σ are also vertices of $\bar{\sigma}$ except the vertex y^i. **Table 9.3.1** shows how $\bar{y}$ and $\bar{\pi}$ depend on y, π and i.

Next, let $\sigma = J_2(y, \pi, s)$ be a simplex of the J_2-triangulation with vertices y^{-1}, y^0, $\cdots$, y^n. Suppose that we want to obtain the simplex of the J_2-triangulation, $\bar{\sigma} = J_2(\bar{y}, \bar{\pi}, \bar{s})$, such that all vertices of σ are also vertices of $\bar{\sigma}$ except the vertex y^i. **Table 9.3.2** shows how $\bar{y}, \bar{\pi}$ and $\bar{s}$ depend on y, π, s and i.

Finally, let a simplex of the D_2-triangulation, $\sigma = D_2(y, \pi, s, p_1, p_2)$, be given with vertices y^{-1}, y^0, $\cdots$, y^n. If we want to obtain a simplex of the D_2-triangulation, $\bar{\sigma} = D_2(\bar{y}, \bar{\pi}, \bar{s}, \bar{p}_1, \bar{p}_2)$, such that all vertices of σ are also vertices of $\bar{\sigma}$ except the vertex y^i, then **Table 9.3.3** shows how $\bar{y}, \bar{\pi}, \bar{s}, \bar{p}_1$ and $\bar{p}_2$ depend on y, π, s, p_1, p_2 and i.

In these tables, $y_0 = 2^{-(k+1)}$, $y = (y_1, y_2, \cdots, y_n)^{\mathsf{T}}$, $\bar{y}_0 = 2^{-(\bar{k}+1)}$, and $\bar{y} = (\bar{y}_1, \bar{y}_2, \cdots, \bar{y}_n)^{\mathsf{T}}$. In addition,

$$i^* = \begin{cases} q - 1 & \text{if } i = q - 2, \\ q - 2 & \text{if } i = q - 1 \end{cases}$$

and

$$i^{**} = \begin{cases} n & \text{if } i = n - 1, \\ n - 1 & \text{if } i = n. \end{cases}$$

Table 9.3.1. The Pivot Rules of the K_2-Triangulation

i	q		$\bar{y}$	$\bar{\pi}$	$\bar{q}$	$\bar{k}$
-1	0		y	$(\pi(1),\ldots,\pi(n),\pi(0))$	n	$k-1$
	$q \ge 1$	$y^0_{\pi(0)} = \alpha_k(w_{\pi(0)}+1)$	$y - (y_{\pi(0)} - \alpha_k w_{\pi(0)})u^{\pi(0)}$	$(\pi(1),\ldots,\pi(n),\pi(0))$	$q-1$	k
		$y^0_{\pi(0)} \ne \alpha_k(w_{\pi(0)}+1)$	$y + \alpha_{k+1}u^{\pi(0)}$	$(\pi(1),\ldots,\pi(q-1),\pi(0),\pi(q),\ldots,\pi(n))$	q	k
$0 \le i < q-1$			y	$(\pi(0),\ldots,\pi(i+1),\pi(i),\ldots,\pi(n))$	q	k
$q-1$	$q \ge 1$	$y_{\pi(q-1)} = \alpha_k(w_{\pi(q-1)}-1)$	y	$(\pi(0),\ldots,\pi(q),\pi(q-1),\ldots,\pi(n))$	$q-1$	k
		$y_{\pi(q-1)} \ne \alpha_k(w_{\pi(q-1)}-1)$	$y - \alpha_{k+1}u^{\pi(q-1)}$	$(\pi(q-1),\pi(0),\ldots,\pi(q-2),\pi(q),\ldots,\pi(n))$	q	k
q	$q < n$	$y^{q+1}_{\pi(q+1)} = \alpha_k(w_{\pi(q+1)}+1)$	y	$(\pi(0),\ldots,\pi(q+1),\pi(q),\ldots,\pi(n))$	$q+1$	k
		$y^{q+1}_{\pi(q+1)} \ne \alpha_k(w_{\pi(q+1)}+1)$	$y + \alpha_k u^{\pi(q+1)}$	$(\pi(0),\ldots,\pi(q),\pi(q+2),\ldots,\pi(n),\pi(q+1))$	q	k
$q < i < n$			y	$(\pi(0),\ldots,\pi(i+1),\pi(i),\ldots,\pi(n))$	q	k
n	$q < n$	$y_{\pi(n)} = \alpha_k(w_{\pi(n)}-1)$	$y + (\alpha_k - \alpha_{k+1})u^{\pi(n)}$	$(\pi(n),\pi(0),\ldots,\pi(n-1))$	$q+1$	k
		$y_{\pi(n)} \ne \alpha_k(w_{\pi(n)}-1)$	$y - \alpha_k u^{\pi(n)}$	$(\pi(0),\ldots,\pi(q),\pi(n),\pi(q+1),\ldots,\pi(n-1))$	q	k
	n		y	$(\pi(n),\pi(0),\ldots,\pi(n-1))$	0	$k+1$

Table 9.3.2. The Pivot Rules of the J_2-Triangulation

i	q		$\bar{y}$	$\bar{s}$	$\bar{\pi}$	$\bar{q}$	$\bar{k}$
-1	0		$y - \alpha_k s$	s	$(\pi(1), \ldots, \pi(n), \pi(0))$	n	$k-1$
-1	$q \geq 1$		$y + 2\alpha_{k+1} s_{\pi(0)} u^{\pi(0)}$	$s - 2s_{\pi(0)} u^{\pi(0)}$	π	q	k
$0 \leq i < q-1$			y	s	$(\pi(0), \ldots, \pi(i+1), \pi(i), \ldots, \pi(n))$	q	k
$q-1$	$q \geq 1$	$y_{\pi(q-1)} = \alpha_k(w_{\pi(q-1)} - s_{\pi(q-1)})$	y	$s - 2s_{\pi(q-1)} u^{\pi(q-1)}$	$(\pi(0), \ldots, \pi(q-2), \pi(q), \ldots, \pi(n), \pi(q-1))$	$q-1$	k
		$y_{\pi(q-1)} \neq \alpha_k(w_{\pi(q-1)} - s_{\pi(q-1)})$	y	$s - 2s_{\pi(q-1)} u^{\pi(q-1)}$	π	q	k
q	$q < n$		y	$s - 2s_{\pi(q+1)} u^{\pi(q+1)}$	π	q	k
$q < i < n$			y	s	$(\pi(0), \ldots, \pi(i+1), \pi(i), \ldots, \pi(n))$	q	k
n	$q < n$		y	$s - 2s_{\pi(n)} u^{\pi(n)}$	$(\pi(0), \ldots, \pi(q-1), \pi(n), \pi(q), \ldots, \pi(n-1))$	$q+1$	k
	n		$y + \alpha_{k+1} s$	s	$(\pi(n), \pi(0), \ldots, \pi(n-1))$	0	$k+1$

Table 9.3.3(1). The Pivot Rules of the D_2-Triangulation

i	q		p_1	p_2	$\bar{y}$	$\bar{s}$	$\bar{\pi}$	$\bar{q}$	$\bar{p}_1$	$\bar{p}_2$	$\bar{k}$
-1	0				$y-\alpha_k s$	s	$(\pi(1),\ldots,\pi(n),\pi(0))$	n	p_2-1	0	k
	1				$y+2\alpha_{k+1}s_{\pi(0)}u^{\pi(0)}$	$s-2s_{\pi(0)}u^{\pi(0)}$	π	q	p_1	p_2	$k-1$
	$q\geq 2$	$y_{\pi(0)}=\alpha_k(w_{\pi(0)}-s_{\pi(0)})$	-1		y	s	π	q	p_1+1	p_2	k
			0		y	s	π	q	p_1-1	p_2	k
			$p_1\geq 1$	0	y	$s-2s_{\pi(0)}u^{\pi(0)}$	$(\pi(1),\ldots,\pi(n),\pi(0))$	$q-1$	p_1-1	p_2	k
				$p_2\geq 1$	y	$s-2s_{\pi(0)}u^{\pi(0)}$	$(\pi(1),\ldots,\pi(q),\pi(0),\pi(q+1),\ldots,\pi(n))$	$q-1$	p_1-1	p_2+1	k
		$y_{\pi(0)}\neq\alpha_k(w_{\pi(0)}-s_{\pi(0)})$			y	$s-2s_{\pi(0)}u^{\pi(0)}$	π	q	p_1	p_2	k
$0\leq i$ $\leq q-1$		$y_{\pi(i)}=\alpha_k(w_{\pi(i)}-s_{\pi(i)})$	-1	0	y	$s-2s_{\pi(i)}u^{\pi(i)}$	$(\pi(0),\ldots,\pi(i-1),\pi(i+1),\ldots,\pi(n),\pi(i))$	$q-1$	p_1	p_2	k
			0	$p_2\geq 1$	y	$s-2s_{\pi(i)}u^{\pi(i)}$	$(\pi(0),\ldots,\pi(i-1),\pi(i+1),\ldots,\pi(q),\pi(i),\pi(q+1),\ldots,\pi(n))$	$q-1$	p_1	p_2+1	k
		$y_{\pi(i)}\neq\alpha_k(w_{\pi(i)}-s_{\pi(i)})$	-1		y	$s-2s_{\pi(i)}u^{\pi(i)}$	π	q	p_1	p_2	k
			$i<p_1$ / -1		y	s	$(\pi(0),\ldots,\pi(i+1),\pi(i),\ldots,\pi(n))$	q	p_1	p_2	k
			$=p_1$ / -1		y	s	π	q	p_1-1	p_2	k
			$>p_1$ / -1 / $0\leq p_1\leq q-2$		y		$(\pi(0),\ldots,\pi(p_1-1),\pi(i),\pi(p_1),\ldots,\pi(i-1),\pi(i+1),\ldots,\pi(n))$	q	p_1+1	p_2	k
			$\geq p_1$ / $0\leq p_1$ / $=q-2$		$y+2\alpha_{k+1}s_{\pi(i^*)}u^{\pi(i^*)}$	$s-2s_{\pi(i^*)}u^{\pi(i^*)}$	π	q	p_1	p_2	k

Table 9.3.3(2). The Pivot Rules of the D_2-Triangulation

i	q	p_1	p_2	$\bar{y}$	$\bar{s}$	$\bar{\pi}$	$\bar{q}$	$\bar{p}_1$	$\bar{p}_2$	$\bar{k}$
q			0	y	s	π	q	p_1	p_2+1	k
	$q \le \frac{n}{2}$	-1	1	y	s	π	q	p_1	p_2-1	k
	$q = \frac{n}{2}-1$		$p_2 \ge 2$	y	s	π	q	p_1	p_2-1	k
		$p_1 \ge 0$		y	$s-2s_{\pi(q+1)}u^{\pi(q+1)}$	$(\pi(0),\ldots,\pi(q+1),\ \pi(q),\ldots,\pi(n))$	$q+1$	p_1+1	p_2	k
				y	$s-2s_{\pi(q+1)}u^{\pi(q+1)}$	$(\pi(q+1),\pi(0),\ldots,\pi(q),\ \pi(q+2),\ldots,\pi(n))$	$q+1$	p_1	p_2	k
$n-1$			0	y	$s-2s_{\pi(q+1)}u^{\pi(q+1)}$	π	q	p_1	p_2	k
	-1			y	$s-2s_{\pi(i)}u^{\pi(i)}$	$(\pi(0),\ldots,\pi(q-1),\pi(i),\ \pi(q),\ldots,\pi(i-1),\ \pi(i+1),\ldots,\pi(n))$	$q+1$	p_1+1	p_2	k
	$p_1 \ge 0$			y	$s-2s_{\pi(i)}u^{\pi(i)}$	$(\pi(i),\pi(0),\ldots,\pi(i-1),\ \pi(i+1),\ldots,\pi(n))$	q	p_1	p_2	k
$q < i \le n$			$i < q+p_2-1$	y	s	$(\pi(0),\ldots,\pi(i+1),\ \pi(i),\ldots,\pi(n))$	q	p_1	p_2-1	k
			$i = q+p_2-1$	y	s	π	q	p_1	p_2+1	k
		$\cdot$	$i > q+p_2-1,$ $1 \le p_2 <$ $n-q-1$	y	s	$(\pi(0),\ldots,\pi(q+p_2-1),\ \pi(i),\pi(q+p_2),\ldots,\pi(i-1),\ \pi(i+1),\ldots,\pi(n))$	q	p_1	p_2	k
			$i \ge p_2+q,$ $1 \le p_2 =$ $n-q-1$	y	$s-2s_{\pi(i^{**})}u^{\pi(i^{**})}$	π	q	p_1	p_2+1	k
n	n			$y+\alpha_{k+1}s$	s	$(\pi(n),\pi(0),\ldots,\pi(n-1))$	0	-1	p_1+1	$k+1$

9.4   Description of the D_2^*-Triangulation

In a similar way as in the first section we can obtain a triangulation of continuous refinement of grid sizes with even factors for $(0, 1] \times R^n$ such that it induces the D_3-triangulation as its special case. This simplicial subdivision is called the D_2^*-triangulation. We describe it as follows. Let N_0 denote the index set $\{0, 1, \cdots, n\}$ and let u^i denote the i-th unit vector in R^{n+1} for $i = 0, 1, \cdots, n$. Let $\beta_{-1} = 1$. For $j = 0, 1, \cdots$, let $\beta_j \in \{1/i \mid i = 1, 2, \cdots\}$ and $\alpha_{j+1} = \alpha_j \beta_j / 2$ with α_0 some positive number in $(0, 1]$.

Take a permutation $\pi = (\pi(0), \pi(1), \cdots, \pi(n))$ of the elements of N_0. Let q denote the integer such that $\pi(q) = 0$. Next, let $K(\pi)$ denote the set

$$\left\{ y \in (0, 1] \times R^n \;\middle|\; \begin{array}{l} \text{for some integer } k \geq 0, \; y_0 = 2^{-(k+1)}, \\ y_{\pi(i)}/2\alpha_{k+1} \text{ is an integer for } i = 0, \cdots, q-1, \\ \text{and } y_{\pi(i)}/\alpha_k \text{ is odd for } i = q+1, \cdots, n \end{array} \right\}.$$

Take $y \in K(\pi)$. Then define

$$w_{\pi(i)} = \begin{cases} \lfloor y_{\pi(i)}/\alpha_k \rfloor + 1 & \text{if } \lfloor y_{\pi(i)}/\alpha_k \rfloor \text{ is odd,} \\ \lfloor y_{\pi(i)}/\alpha_k \rfloor & \text{otherwise,} \end{cases}$$

for $i = 0, 1, \cdots, q-1$.

Definition 9.4.1. Let y and π be given as above. Then y^{-1}, y^0, $\cdots$, y^n are given as follows:

$$\begin{aligned} y^{-1} &= y, \\ y^i &= y^{i-1} + 2\alpha_{k+1} u^{\pi(i)}, \; i = 0, 1, \cdots, q-1, \\ y^q &= \alpha_k \sum_{j=0}^{q-1} w_{\pi(j)} u^{\pi(j)} + \sum_{j=q+1}^{n} (y_{\pi(j)} - \alpha_k) u^{\pi(j)} + 2y_0 u^0, \\ y^i &= y^{i-1} + 2\alpha_k u^{\pi(i)}, \; i = q+1, \cdots, n. \end{aligned}$$

Let y^{-1}, y^0, $\cdots$, y^n be obtained from the above definition. Then it is obvious that they are affinely independent. Thus their convex hull is a simplex. Let us denote this simplex by $K_2^*(y, \pi)$. Let K_2^* denote the collection of all such simplices $K_2^*(y, \pi)$. Then following the conclusions in the first section, we have that K_2^* is a grid refinement triangulation of $(0, 1] \times R^n$ such that its factor of refinement can be chosen as any even integer. We call it the K_2^*-triangulation. When the factor of refinement is always chosen equal to two, the K_3-triangulation is induced as its special case.

Take a permutation $\pi = (\pi(0), \pi(1), \cdots, \pi(n))$ of the elements of N_0. Let q denote the integer such that $\pi(q) = 0$. Next, let $J(\pi)$ denote the set

$$
\left\{ y \in (0,1] \times R^n \left|
\begin{array}{l}
\text{for some integer } k \geq 0,\ y_0 = 2^{-(k+1)}, \\
\text{if } 1/\beta_k \text{ is even, } y_{\pi(i)}/2\alpha_{k+1} \text{ is even} \\
\text{for } i = 0, \cdots, q-1, \\
\text{if } 1/\beta_k \text{ is odd, } y_{\pi(i)}/2\alpha_{k+1} \text{ is odd} \\
\text{for } i = 0, \cdots, q-1, \\
\text{and } y_{\pi(i)}/\alpha_k \text{ is odd for } i = q+1, \cdots, n
\end{array}
\right. \right\}.
$$

Take $y \in J(\pi)$. Then define

$$
w_{\pi(i)} = \left\{
\begin{array}{ll}
\lfloor y_{\pi(i)}/\alpha_k \rfloor + 1 & \text{if } \lfloor y_{\pi(i)}/\alpha_k \rfloor \text{ is odd and} \\
& \quad \text{either } y_{\pi(i)}/\alpha_k \neq \lfloor y_{\pi(i)}/\alpha_k \rfloor \\
& \quad \text{or both } \lfloor y_{\pi(i)}/\alpha_k \rfloor = y_{\pi(i)}/\alpha_k \text{ and } s_{\pi(i)} = 1, \\
\lfloor y_{\pi(i)}/\alpha_k \rfloor & \text{if } \lfloor y_{\pi(i)}/\alpha_k \rfloor \text{ is even,} \\
\lfloor y_{\pi(i)}/\alpha_k \rfloor - 1 & \text{otherwise,}
\end{array}
\right.
$$

for $i = 0, 1, \cdots, q-1$. If $1/\beta_{k-1}$ is odd, let

$$
t_{\pi(i)} = \left\{
\begin{array}{ll}
-1 & \text{if } y_{\pi(i)}/\alpha_k = 1 (\mathrm{mod}4), \\
1 & \text{if } y_{\pi(i)}/\alpha_k = 3 (\mathrm{mod}4),
\end{array}
\right.
$$

for $i = q+1, \cdots, n$ and if $1/\beta_{k-1}$ is even, let

$$
t_{\pi(i)} = \left\{
\begin{array}{ll}
1 & \text{if } y_{\pi(i)}/\alpha_k = 1 (\mathrm{mod}4), \\
-1 & \text{if } y_{\pi(i)}/\alpha_k = 3 (\mathrm{mod}4),
\end{array}
\right.
$$

for $i = q+1, \cdots, n$. Take a sign vector $s = (s_1, s_2, \cdots, s_n)^\top$ such that $s_i \in \{-1, +1\}$ for $i = 1, 2, \cdots, n$ and $s_{\pi(i)} = t_{\pi(i)}$ for $i = q+1, \cdots, n$.

Definition 9.4.2. Let y, π and s be given as above. Then y^{-1}, y^0, $\cdots$, y^n are given as follows:

$$
\begin{array}{ll}
y^{-1} & = y, \\
y^i & = y^{i-1} + 2\alpha_{k+1}s_{\pi(i)}u^{\pi(i)}, \ i = 0, 1, \cdots, q-1, \\
y^q & = \alpha_k \sum_{j=0}^{q-1} w_{\pi(j)}u^{\pi(j)} + \sum_{j=q+1}^{n}(y_{\pi(j)} - \alpha_k s_{\pi(j)})u^{\pi(j)} + 2y_0 u^0, \\
y^i & = y^{i-1} + 2\alpha_k s_{\pi(i)}u^{\pi(i)}, \ i = q+1, \cdots, n.
\end{array}
$$

Let $y^{-1}, y^0, \cdots, y^n$ be obtained in the above manner. Then it is obvious that they are affinely independent. Thus their convex hull is a simplex. Let us denote this simplex by $J_2^*(y, \pi, s)$. Let J_2^* denote the set of all such simplices $J_2^*(y, \pi, s)$. Then following the conclusions in the first section, we have that J_2^* is a grid refinement triangulation of $(0, 1] \times R^n$ such that its factor of refinement can be chosen as any even integer. We call it the J_2^*-triangulation. When the factor of refinement is always chosen equal to two, the J_3-triangulation is induced as its special case.

Take a permutation $\pi = (\pi(0), \pi(1), \cdots, \pi(n))$ of the elements of N_0. Let q denote the integer such that $\pi(q) = 0$. Next, let $D(\pi)$ denote the set

$$
\left\{
y \in (0, 1] \times R^n
\;\middle|\;
\begin{array}{l}
\text{for some integer } k \geq 0,\; y_0 = 2^{-(k+1)}, \\
\text{if } 1/\beta_k \text{ is even then } y_{\pi(i)}/2\alpha_{k+1} \text{ is even} \\
\text{for } i = 0, \cdots, q-1, \\
\text{if } 1/\beta_k \text{ is odd then } y_{\pi(i)}/2\alpha_{k+1} \text{ is odd} \\
\text{for } i = 0, \cdots, q-1, \\
\text{and } y_{\pi(i)}/\alpha_k \text{ is odd for } i = q+1, \cdots, n
\end{array}
\right\}
$$

Take $y \in D(\pi)$. Then define

$$
w_{\pi(i)} =
\begin{cases}
\lfloor y_{\pi(i)}/\alpha_k \rfloor + 1 & \text{if } \lfloor y_{\pi(i)}/\alpha_k \rfloor \text{ is odd and} \\
& \quad \text{either } y_{\pi(i)}/\alpha_k \neq \lfloor y_{\pi(i)}/\alpha_k \rfloor \\
& \quad \text{or both } \lfloor y_{\pi(i)}/\alpha_k \rfloor = y_{\pi(i)}/\alpha_k \text{ and } s_{\pi(i)} = 1, \\
\lfloor y_{\pi(i)}/\alpha_k \rfloor & \text{if } \lfloor y_{\pi(i)}/\alpha_k \rfloor \text{ is even,} \\
\lfloor y_{\pi(i)}/\alpha_k \rfloor - 1 & \text{otherwise,}
\end{cases}
$$

for $i = 0, 1, \cdots, q-1$. If $1/\beta_{k-1}$ is odd, let

$$
t_{\pi(i)} =
\begin{cases}
-1 & \text{if } y_{\pi(i)}/\alpha_k = 1(\mathrm{mod}4), \\
1 & \text{if } y_{\pi(i)}/\alpha_k = 3(\mathrm{mod}4),
\end{cases}
$$

for $i = q+1, \cdots, n$ and if $1/\beta_{k-1}$ is even, let

$$
t_{\pi(i)} =
\begin{cases}
1 & \text{if } y_{\pi(i)}/\alpha_k = 1(\mathrm{mod}4), \\
-1 & \text{if } y_{\pi(i)}/\alpha_k = 3(\mathrm{mod}4),
\end{cases}
$$

for $i = q+1, \cdots, n$. Take a sign vector $s = (s_1, s_2, \cdots, s_n)^\top$ such that $s_i \in \{-1, +1\}$ for $i = 1, 2, \cdots, n$ and $s_{\pi(i)} = t_{\pi(i)}$ for $i = q+1, \cdots, n$.

Then define

$$
I = \begin{cases} \left\{ \pi(i) \mid w_{\pi(i)}/2 \text{ is even and } 0 \le i \le q - 1 \right\} & \text{if } 1/\beta_{k-1} \text{ is odd,} \\[2mm] \left\{ \pi(i) \mid w_{\pi(i)}/2 \text{ is odd and } 0 \le i \le q - 1 \right\} & \text{if } 1/\beta_{k-1} \text{ is even} \end{cases}
$$

and let h denote the number of elements in I. Take two integers p_1 and p_2 such that $-1 \le p_1 \le q - 2$, if $h = 0$ then $0 \le p_2 \le n - q - 1$, and if $h > 0$ then $p_2 = n - q$.

Definition 9.4.3. Let y, π, s, p_1 and p_2 be given as above. Then y^{-1}, y^0, $\cdots$, y^n are given as follows.

When $p_1 = -1$, $y^{-1} = y$,

$$
y^i = y + 2\alpha_{k+1} s_{\pi(i)} u^{\pi(i)}, \quad i = 0, 1, \cdots, q - 1,
$$

and when $p_1 \ge 0$,

$$
\begin{aligned}
y^{-1} &= y + 2\alpha_{k+1} \sum_{j=0}^{q-1} s_{\pi(j)} u^{\pi(j)}, \\
y^i &= y^{i-1} - 2\alpha_{k+1} s_{\pi(i)} u^{\pi(i)}, \quad i = 0, 1, \cdots, p_1 - 1, \\
y^i &= y + 2\alpha_{k+1} s_{\pi(i)} u^{\pi(i)}, \quad i = p_1, \cdots, q - 1.
\end{aligned}
$$

When $h > 0$,

$$
\begin{aligned}
y^q &= \alpha_k \sum_{j=0}^{q-1} w_{\pi(j)} u^{\pi(j)} + \sum_{j=q+1}^{n} (y_{\pi(j)} + \alpha_k s_{\pi(j)}) u^{\pi(j)} + 2y_0 u^0, \\
y^i &= y^{i-1} - 2\alpha_k s_{\pi(i)} u^{\pi(i)}, \quad i = q + 1, \cdots, n,
\end{aligned}
$$

and when $h = 0$, if $p_2 = 0$, then

$$
\begin{aligned}
y^q &= \alpha_k \sum_{j=0}^{q-1} w_{\pi(j)} u^{\pi(j)} + \sum_{j=q+1}^{n} (y_{\pi(j)} - \alpha_k s_{\pi(j)}) u^{\pi(j)} + 2y_0 u^0, \\
y^i &= y^q + 2\alpha_k s_{\pi(i)} u^{\pi(i)}, \quad i = q + 1, \cdots, n,
\end{aligned}
$$

and if $p_2 \ge 1$, then

$$
\begin{aligned}
y^q &= \alpha_k \sum_{j=0}^{q-1} w_{\pi(j)} u^{\pi(j)} + \sum_{j=q+1}^{n} (y_{\pi(j)} + \alpha_k s_{\pi(j)}) u^{\pi(j)} + 2y_0 u^0, \\
y^i &= y^{i-1} - 2\alpha_k s_{\pi(i)} u^{\pi(i)}, \quad i = q + 1, \cdots, q + p_2 - 1, \\
y^i &= y^* + 2\alpha_k s_{\pi(i)} u^{\pi(i)}, \quad i = q + p_2, \cdots, n,
\end{aligned}
$$

where

$$y^* = \alpha_k \sum_{j=0}^{q-1} w_{\pi(j)} u^{\pi(j)} + \sum_{j=q+1}^{n} (y_{\pi(j)} - \alpha_k s_{\pi(j)}) u^{\pi(j)} + 2y_0 u^0.$$

Let y^{-1}, y^0, $\cdots$, y^n be obtained from the above definition. Then it is obvious that they are affinely independent. Thus their convex hull is a simplex. Let us denote this simplex by $D_2^*(y, \pi, s, p_1, p_2)$. Let D_2^* denote the set of all such simplices $D_2^*(y, \pi, s, p_1, p_2)$. Then following the conclusions in the first section, we have that D_2^* is a grid refinement triangulation of $(0, 1] \times R^n$ such that its factor of refinement can be chosen as any even integer. We call it the D_2^*-triangulation. When the factor of refinement is always chosen equal to two, the D_3-triangulation is induced as its special case.

9.5   Pivot Rules of the D_2^*-Triangulation

A simplicial homotopy algorithm, based on one of the K_2^*-triangulation, the J_2^*-triangulation, or the D_2^*-triangulation, can be implemented according to the following pivot rules.

Let a simplex of the K_2^*-triangulation, $\sigma = K_2^*(y, \pi)$, be given with vertices y^{-1}, y^0, $\cdots$, y^n. We wish to obtain the simplex of the K_2^*-triangulation, $\bar{\sigma} = K_2^*(\bar{y}, \bar{\pi})$, such that all vertices of σ are also vertices of $\bar{\sigma}$ except the vertex y^i. **Table 9.5.1** shows how $\bar{y}$ and $\bar{\pi}$ depend on y, π and i.

Next, let a simplex of the J_2^*-triangulation, $\sigma = J_2^*(y, \pi, s)$, be given with vertices y^{-1}, y^0, $\cdots$, y^n. We wish to obtain the simplex of the J_2^*-triangulation, $\bar{\sigma} = J_2^*(\bar{y}, \bar{\pi}, \bar{s})$, such that all vertices of σ are also vertices of $\bar{\sigma}$ except the vertex y^i. **Table 9.5.2** shows how $\bar{y}, \bar{\pi}$ and $\bar{s}$ depend on y, π, s and i.

In addition, let a simplex of the D_2^*-triangulation, $\sigma = D_2^*(y, \pi, s, p_1, p_2)$, be given with vertices y^{-1}, y^0, $\cdots$, y^n. We wish to obtain the simplex of the D_2^*-triangulation, $\bar{\sigma} = D_2^*(\bar{y}, \bar{\pi}, \bar{s}, \bar{p}_1, \bar{p}_2)$, such that all vertices of σ are also vertices of $\bar{\sigma}$ except the vertex y^i. **Table 9.5.3** shows how $\bar{y}, \bar{\pi}, \bar{s}, \bar{p}_1$ and $\bar{p}_2$ depend on y, π, s, p_1, p_2 and i.

In these tables, $y_0 = 2^{-(k+1)}$, $y = (y_1, \cdots, y_n)^{\top}$, $\bar{y}_0 = 2^{-(\bar{k}+1)}$, and

$\bar{y} = (\bar{y}_1, \cdots, \bar{y}_n)^\mathsf{T}$. Moreover, $u = (1, 1, \cdots, 1)^\mathsf{T}$. In addition,

$$i^* = \begin{cases} q - 1 & \text{if } i = q - 2, \\ q - 2 & \text{if } i = q - 1 \end{cases}$$

and

$$i^{**} = \begin{cases} n & \text{if } i = n - 1, \\ n - 1 & \text{if } i = n. \end{cases}$$

Table 9.5.1. The Pivot Rules of the K_2^*-Triangulation

i	q		$\bar{y}$	$\bar{\pi}$	$\bar{q}$	$\bar{k}$
-1	0		$y - \alpha_k u$	$(\pi(1),\ldots,\pi(n),\pi(0))$	n	$k-1$
	$q \ge 1$	$y^0_{\pi(0)} = \alpha_k(w_{\pi(0)}+1)$	$y - (y_{\pi(0)} - \alpha_k(w_{\pi(0)}+1))u^{\pi(0)}$	$(\pi(1),\ldots,\pi(n),\pi(0))$	$q-1$	k
		$y^0_{\pi(0)} \neq \alpha_k(w_{\pi(0)}+1)$	$y + 2\alpha_{k+1}u^{\pi(0)}$	$(\pi(1),\ldots,\pi(q-1),$ $\pi(0),\pi(q),\ldots,\pi(n))$	q	k
$0 \leq i < q-1$			y	$(\pi(0),\ldots,\pi(i+1),$ $\pi(i),\ldots,\pi(n))$	q	k
$q-1$	$q > 0$	$y_{\pi(q-1)} = \alpha_k(w_{\pi(q-1)}-1)$	y	$(\pi(0),\ldots,\pi(q),$ $\pi(q-1),\ldots,\pi(n))$	$q-1$	k
		$y_{\pi(q-1)} \neq \alpha_k(w_{\pi(q-1)}-1)$	$y - 2\alpha_{k+1}u^{\pi(q-1)}$	$(\pi(q-1),\pi(0),\ldots,\pi(q-2),$ $\pi(q),\ldots,\pi(n))$	q	k
q	$q < n$		y	$(\pi(0),\ldots,\pi(q+1),$ $\pi(q),\ldots,\pi(n))$	$q+1$	k
$q < i < n$			y	$(\pi(0),\ldots,\pi(i+1),$ $\pi(i),\ldots,\pi(n))$	q	k
n	$q < n$		$y - 2\alpha_{k+1}u^{\pi(n)}$	$(\pi(n),\pi(0),\ldots,\pi(n-1))$	$q+1$	k
	n		$y + \alpha_{k+1}u$	$(\pi(n),\pi(0),\ldots,\pi(n-1))$	0	$k+1$

Table 9.5.2. The Pivot Rules of the J_2^*-Trianulation

i	q	condition	$\bar{y}$	$\bar{s}$	$\bar{\pi}$	$\bar{q}$	$\bar{k}$
-1	0		$y - \alpha_k s$	s	$(\pi(1), \pi(2), \ldots, \pi(n), \pi(0))$	n	$k-1$
	$q > 0$		$y + 4\alpha_{k+1} s_{\pi(0)} u^{\pi(0)}$	$s - 2s_{\pi(0)} u^{\pi(0)}$	π	q	k
$0 \leq i < q-1$			y	s	$(\pi(0), \ldots, \pi(i+1), \pi(i), \ldots, \pi(n))$	q	k
$q-1$	$q > 0$	$y_{\pi(q-1)} = \alpha_k(w_{\pi(q-1)} - s_{\pi(q-1)})$; $\;s_{\pi(q-1)} = t_{\pi(q-1)}$	y	s	$(\pi(0), \ldots, \pi(q), \pi(q-1), \ldots, \pi(n))$	$q-1$	k
		$s_{\pi(q-1)} \neq t_{\pi(q-1)}$	y	$s - 2s_{\pi(q-1)} u^{\pi(q-1)}$	$(\pi(0), \ldots, \pi(q-2), \pi(q), \pi(n), \pi(q-1))$	$q-1$	k
		$y_{\pi(q-1)} \neq \alpha_k(w_{\pi(q-1)} - s_{\pi(q-1)})$.	y	$s - 2s_{\pi(q-1)} u^{\pi(q-1)}$	π	q	k
q	$q < n$		y	s	$(\pi(0), \ldots, \pi(q+1), \pi(q), \ldots, \pi(n))$	$q+1$	k
$q < i < n$			y	s	$(\pi(0), \ldots, \pi(i+1), \pi(i), \ldots, \pi(n))$	q	k
n	$q < n$		y	$s - 2s_{\pi(n)} u^{\pi(n)}$	$(\pi(0), \ldots, \pi(q-1), \pi(q), \ldots, \pi(n-1))$	$q+1$	k
	n		$y + \alpha_{k+1} s$	s	$(\pi(n), \pi(0), \ldots, \pi(n-1))$	0	$k+1$

Table 9.5.3(1). The Pivot Rules of the D_2^*-Triangulation

i	q		p_1			p_2	$\bar{y}$	$\bar{s}$	$\bar{\pi}$	$\bar{p}_1$	$\bar{p}_2$	$\bar{q}$	$\bar{k}$
-1	0						$y-\alpha_k s$	s	$(\pi(1),\ldots,\pi(n),\pi(0))$	p_2-1	0	n	$k-1$
	1						$y+4\alpha_{k+1}s_{\pi(0)}u^{\pi(0)}$	$s-2s_{\pi(0)}u^{\pi(0)}$	π	p_1	p_2	q	k
	$q>1$		-1				y	s	π	p_1+1	p_2	q	k
		$y_{\pi(0)}=\alpha_k(w_{\pi(0)}-s_{\pi(0)})$	0	$h=0$	$p_2=0$		y	$s-2s_{\pi(0)}u^{\pi(0)}$	π	p_1-1	p_2	q	k
			$p_1\ge1$		$p_2\ge1$		y	$s-2s_{\pi(0)}u^{\pi(0)}$	$(\pi(1),\ldots,\pi(q),\pi(0),\pi(q+1),\ldots,\pi(n))$	p_1-1	p_2+1	$q-1$	k
		$s_{\pi(0)}=t_{\pi(0)}$		$h>0$	$h=1$		y	s	$(\pi(1),\ldots,\pi(n),\pi(0))$	p_1-1	p_2	$q-1$	k
		$s_{\pi(0)}\neq t_{\pi(0)}$			$h>1$		y	s	$(\pi(1),\ldots,\pi(n),\pi(0))$	p_1-1	p_2+1	$q-1$	k
		$y_{\pi(0)}\neq\alpha_k(w_{\pi(0)}-s_{\pi(0)})$					y	$s-2s_{\pi(0)}u^{\pi(0)}$	$(\pi(1),\ldots,\pi(q),\pi(0),\pi(q+1),\ldots,\pi(n))$	p_1-1	p_2+1	$q-1$	k
							y	$s-2s_{\pi(0)}u^{\pi(0)}$	π	p_1	p_2	q	k
$0\le i$ $<q$		$y_{\pi(i)}=\alpha_k(w_{\pi(i)}-s_{\pi(i)})$	-1	$h=0$	$p_2=0$		y	$s-2s_{\pi(i)}u^{\pi(i)}$	$(\pi(0),\ldots,\pi(i-1),\pi(i+1),\ldots,\pi(n),\pi(i))$	p_1	p_2	$q-1$	k
			0		$p_2\ge1$		y	$s-2s_{\pi(i)}u^{\pi(i)}$	$(\pi(0),\ldots,\pi(i+1),\ldots,\pi(q),\pi(i),\pi(q+1),\ldots,\pi(n))$	p_1	p_2+1	$q-1$	k
		$s_{\pi(i)}=t_{\pi(i)}$		$h>0$	$h=1$		y	s	$(\pi(0),\ldots,\pi(i-1),\pi(i+1),\ldots,\pi(n),\pi(i))$	p_1	p_2	$q-1$	k
		$s_{\pi(i)}\neq t_{\pi(i)}$			$h>1$		y	s	$(\pi(0),\ldots,\pi(i-1),\pi(i+1),\ldots,\pi(n),\pi(i))$	p_1	p_2+1	$q-1$	k
							y	$s-2s_{\pi(i)}u^{\pi(i)}$	$(\pi(0),\ldots,\pi(i+1),\ldots,\pi(q),\pi(i),\pi(q+1),\ldots,\pi(n))$	p_1	p_2+1	$q-1$	k

Table 9.5.3(2). The Pivot Rules of the D_2^*-Triangulation

i	q	$y_{\pi(i)} \neq \alpha_k(w_{\pi(i)}) - s_{\pi(i)}$	p_1		p_2	$\bar{y}$	$\bar{s}$	$\bar{\pi}$	$\bar{p}_1$	$\bar{p}_2$	$\bar{q}$	$\bar{k}$
$0 \leq i < q$			-1			y	$s - 2s_{\pi(i)}$, $u^{\pi(i)}$	π	p_1	p_2	q	k
			$i < p_1 - 1$			y	s	$(\pi(0),\ldots,\pi(i+1),$ $\pi(i),\ldots,\pi(n))$	p_1	p_2	q	k
			$i = p_1 - 1$				s	π	$p_1 - 1$	p_2	q	k
			$i \geq p_1$				s	$(\pi(0),\ldots,\pi(p_1-1),$ $\pi(i),\pi(p_1),\ldots,\pi(i-1),$ $\pi(i+1),\ldots,\pi(n))$	$p_1 + 1$	p_2	q	k
			$0 \leq p_1 < q-2$ $i \geq q-2$ $0 \leq p_1 = q-2$			$y + 4\alpha_{k+1}$ $s_{\pi(i^*)}u^{\pi(i^*)}$	$s - 2s_{\pi(i^*)}$, $u^{\pi(i^*)}$	π	p_1	p_2		k
q	$q < n-1$		-1	$h = 0$	0	y	s	π	p_1	p_2	q	k
	$n-1$					y	s	$(\pi(0),\ldots,\pi(q+1),$ $\pi(q),\ldots,\pi(n))$	p_1	$p_2 + 1$	q	k
			$p_1 \geq 0$			y	s	$(\pi(q+1),\pi(0),\ldots,\pi(q),$ $\pi(q+2),\ldots,\pi(n))$	$p_1 + 1$	p_2	$q+1$	k
			-1		1	y	$s - 2s_{\pi(q+1)}$, $u^{\pi(q+1)}$	$(\pi(0),\ldots,\pi(q+1),$ $\pi(q),\ldots,\pi(n))$	p_1	$p_2 - 1$	$q+1$	k
					$p_2 \geq 2$	y	$s - 2s_{\pi(q+1)}$, $u^{\pi(q+1)}$	$(\pi(q+1),\pi(0),\ldots,\pi(q),$ $\pi(q+2),\ldots,\pi(n))$	p_1	$p_2 - 1$	$q+1$	k
	$q < n$		$p_1 \geq 0$	$h > 0$		y	$s - 2s_{\pi(q+1)}$, $u^{\pi(q+1)}$	$(\pi(q+1),\pi(0),\ldots,\pi(q),$ $\pi(q+2),\ldots,\pi(n))$	$p_1 + 1$	$p_2 - 1$	$q+1$	k
			-1			y	$s - 2s_{\pi(q+1)}$, $u^{\pi(q+1)}$	$(\pi(0),\ldots,\pi(q+1),$ $\pi(q),\ldots,\pi(n))$	p_1	$p_2 - 1$	$q+1$	k
			$p_1 \geq 0$			y	$s - 2s_{\pi(q+1)}$, $u^{\pi(q+1)}$	$(\pi(q+1),\pi(0),\ldots,\pi(q),$ $\pi(q+2),\ldots,\pi(n))$	$p_1 + 1$	$p_2 - 1$	$q+1$	k
n						$y + \alpha_{k+1}s$	s	$(\pi(n),\pi(0),\ldots,\pi(n-1))$	-1	$p_1 + 1$	0	$k+1$

Table 9.5.3(3). The Pivot Rules of the D_2^*-Triangulation

i	q	p_1	h	p_2	$\bar y$	$\bar s$	$\bar\pi$	$\bar p_1$	$\bar p_2$	$\bar q$	k
$q \le i \le n$		-1	$h=0$	0	y	$s - 2s_{\pi(i)}u^{\pi(i)}$	$(\pi(0),\ldots,\pi(q-1),\pi(i),$ $\pi(q),\ldots,\pi(i-1),$ $\pi(i+1),\ldots,\pi(n))$	p_1	p_2	$q+1$	k
		$p_1 \ge 0$		$i < q$ $+p_2-1$	y	$s - 2s_{\pi(i)}u^{\pi(i)}$	$(\pi(i),\pi(0),\ldots,\pi(i-1),$ $\pi(i+1),\ldots,\pi(n))$	p_1+1	p_2	$q+1$	k
				$i = q$ $+p_2-1$	y	s	$(\pi(0),\ldots,\pi(i+1),$ $\pi(i),\ldots,\pi(n))$	p_1	p_2	q	k
				$i \ge q+p_2$ $1 \le p_2 <$ $n-q-1$	y	s	π	p_1	p_2-1	q	k
				$i \ge n-1$ $1 \le p_2$ $= n-q-1$	y	s	$(\pi(0),\ldots,\pi(q+p_2-1),$ $\pi(i),\pi(q+p_2),\ldots,\pi(i-1),$ $\pi(i+1),\ldots,\pi(n))$	p_1	p_2+1	q	k
		-1	$h>0$		y	s	$(\pi(0),\ldots,\pi(q-1),\pi(i^{**}),$ $\pi(q),\ldots,\pi(i^{**}-1),$ $\pi(i^{**}+1),\ldots,\pi(n))$	p_1	p_2	$q+1$	k
$q < i$ $< n$		$p_1 \ge 0$			y	s	$(\pi(i^{**}),\pi(0),\ldots,\pi(i^{**}-1),$ $\pi(i^{**}+1),\ldots,\pi(n))$	p_1+1	p_2	$q+1$	k
		-1			y	s	$(\pi(0),\ldots,\pi(i+1),$ $\pi(i),\ldots,\pi(n))$	p_1	p_2	q	k
	$q<n$	$p_1 \ge 0$			y	s	$(\pi(0),\ldots,\pi(q-1),\pi(n),$ $\pi(q),\ldots,\pi(n-1))$	p_1	p_2-1	$q+1$	k
n					y	s	$(\pi(n),\pi(0),\ldots,\pi(n-1))$	p_1+1	p_2-1	$q+1$	k

9.6 Comparison of Several Triangulations for Simplicial Homotopy Algorithms

In this section we first calculate the number of simplices of the K_2-triangulation, the J_2-triangulation and the D_2-triangulation in the set $[1/2, 1] \times H^n$, where H^n denotes the set

$$\{x \in R^n \mid 0 \le x_i \le 2 \text{ for } i = 1, 2, \cdots, n\}.$$

For simplicity, we take $\alpha_0 = 1$ and set $\alpha = 1/\beta_0$.

Theorem 9.6.1. The number of simplices of both the K_2-triangulation and the J_2-triangulation in the set $[1/2, 1] \times H^n$ is equal to

$$p_n(\alpha) = \begin{cases} (1 - \alpha^{n+1})2^n n!/(1 - \alpha) & \text{if } \alpha \ne 1 \\ (n + 1)2^n n! & \text{if } \alpha = 1. \end{cases}$$

The number of simplices of the D_2-triangulation in the set $[1/2, 1] \times H^n$ is equal to

$$q_n(\alpha) = 2^n \sum_{m=0}^{n} \alpha^m C_n^m d_m d_{n-m}$$

where

$$d_j = j + j(j - 1) + \cdots + j(j - 1) \cdots 4 \cdot 3 + 2$$

for $j \ge 2$, $d_0 = d_1 = 1$, and $C_n^m = \frac{n!}{m!(n-m)!}$.

Proof. Let $\bar{Q}$ denote the set $\{w \in R^n \mid w_i \in \{0, 1, 2\} \text{ for } i = 1, 2, \cdots, n\}$. Take $w \in \bar{Q}$. Let $\bar{A}(w)$ denote the set

$$\{x \in R^n \mid w_i - 1 \le x_i \le w_i + 1 \text{ for } i \in I_o(w) \text{ and } x_i = w_i \text{ for } i \in I_e(w)\}$$

and let $\bar{B}(w)$ denote the set

$$\left\{ x \in R^n \middle| \begin{array}{l} x_i = w_i \text{ for } i \in I_o(w), \\ w_i \le x_i \le w_i + 1 \text{ for } i \in I_e(w) \text{ and } w_i = 0, \\ w_i - 1 \le x_i \le w_i \text{ for } i \in I_e(w) \text{ and } w_i = 2 \end{array} \right\}.$$

Next let $\bar{D}(w)$ denote the convex hull of the set

$$(\{1\} \times \bar{A}(w)) \cup (\{1/2\} \times \bar{B}(w)).$$

Then it is obvious that

$$[1/2, 1] \times H^n = \cup_{w \in \bar{Q}} \bar{D}(w).$$

Let m denote the number of elements in $I_e(w)$. Then there are $2^m C_n^m$ elements in $\bar{Q}$ such that m components of each of them are even. Thus the number of simplices of both the K_2-triangulation and the J_2-triangulation in the set $\cup_{w \in \bar{Q}, |I_e(w)|=m} \bar{D}(w)$ is equal to

$$2^m 2^{n-m} \alpha^m C_n^m m!(n-m)!.$$

The number of simplices of the D_2-triangulation in the same set is equal to

$$2^m 2^{n-m} \alpha^m C_n^m d_m d_{n-m}.$$

Since

$$\cup_{m=0}^n (\cup_{w \in \bar{Q}, |I_e(w)|=m} \bar{D}(w)) = [1/2, 1] \times H^n,$$

the theorem follows immediately.

END

Theorem 9.6.2. When $n \geq 3$, $q_n(\alpha) < p_n(\alpha)$. As n goes to infinity, $q_n(\alpha)/p_n(\alpha)$ converges to some number smaller than or equal to $e - 2$.

Proof. The conclusion is obvious, the proof is omitted.

END

From **Theorem 9.6.2**, we have that the number of simplices of the D_2-triangulation is the smallest one for the G_2-triangulations for simplicial homotopy algorithms. Furthermore, since the D_2-triangulation subdivides each level $\{2^{-k}\} \times R^n$, for $k = 0, 1, \cdots$, into simplices according to the D_1-triangulation of R^n, the average directional density of the D_2-triangulation is on each level less than the ones of the K_2-triangulation and of the J_2-triangulation which triangulate each level $\{2^{-k}\} \times R^n$, for $k = 0, 1, \cdots$, according to the K_1-triangulation of R^n and the J_1-triangulation of R^n, respectively.

Next, we calculate the number of simplices of the K_2^*-triangulation, the J_2^*-triangulation and the D_2^*-triangulation in the set $[1/2, 1] \times 2\alpha_0 U^n$, where U^n is the n-dimensional unit cube. Set $\alpha = 1/\beta_0$.

Theorem 9.6.3. The number of simplices of both the K_2^*-triangulation and of the J_2^*-triangulation in the set $[1/2, 1] \times 2\alpha_0 U^n$ is equal to

$$p_n^*(\alpha) = ((2\alpha)^{n+1} - 1)n!/(2\alpha - 1).$$

The number of simplices of the D_2^*-triangulation in the same set is equal to

$$q_n^*(\alpha) = \sum_{m=0}^{n} ((2^m - 1)C_n^m \alpha^m d_m (n - m)! + C_n^m \alpha^m d_m d_{n-m}),$$

where

$$d_j = j + j(j - 1) + \cdots + j(j - 1) \cdots 4 \cdot 3 + 2$$

for $j \geq 2$, $d_0 = d_1 = 1$, and $C_n^m = \frac{n!}{m!(n-m)!}$.

Proof. Let $\bar{Q}$ denote the set $\{w \in R^n \mid w_i \in \{0, 1, 2\} \text{ for } i = 1, 2, \cdots, n\}$. Take $w \in \bar{Q}$. Let $\bar{A}(w)$ denote the set

$$\{x \in R^n \mid w_i - 1 \leq x_i \leq w_i + 1 \text{ for } i \in I_o(w) \text{ and } x_i = w_i \text{ for } i \in I_e(w)\}$$

and let $\bar{B}(w)$ denote the set

$$\left\{ x \in R^n \middle| \begin{array}{l} x_i = w_i \text{ for } i \in I_o(w), \\ w_i \leq x_i \leq w_i + 1 \text{ for } i \in I_e(w) \text{ and } w_i = 0, \\ w_i - 1 \leq x_i \leq w_i \text{ for } i \in I_e(w) \text{ and } w_i = 2 \end{array} \right\}.$$

Furthermore, let $\alpha_0 \bar{D}(w)$ denote the convex hull of the set $(\{1\} \times \alpha_0 \bar{A}(w)) \cup (\{1/2\} \times \alpha_0 \bar{B}(w))$. Then it is obvious that

$$[1/2, 1] \times 2\alpha_0 U^n = \cup_{w \in \bar{Q}} \alpha_0 \bar{D}(w).$$

Let m denote the number of elements in $I_e(w)$. Then there are $2^m C_n^m$ elements in $\bar{Q}$ such that m components of each of them are even. Thus the number of simplices of both the K_2^*-triangulation and the J_2^*-triangulation in the set

$$\cup_{w \in \bar{Q}, |I_e(w)|=m} \alpha_0 \bar{D}(w)$$

is equal to

$$2^m \alpha^m C_n^m (n - m)! m! = (2\alpha)^m n!.$$

The number of simplices of the D_2^*-triangulation in the same set is equal to

$$(2^m - 1)C_n^m \alpha^m d_m (n - m)! + C_n^m \alpha^m d_m d_{n-m}.$$

Since

$$\cup_{m=0}^n (\cup_{w \in \bar{Q}, |I_e(w)|=m} \alpha_0 \bar{D}(w)) = [1/2, 1] \times 2\alpha_0 U^n,$$

the theorem follows immediately.

END

Theorem 9.6.4. When $n \geq 3$, $q_n^*(\alpha) < p_n^*(\alpha)$. And as n goes to infinity, $q_n^*(\alpha)/p_n^*(\alpha)$ converges to $e - 2$.

Proof. The conclusion is obvious, the proof is omitted.

END

From **Theorem 9.6.4**, we have that the number of simplices of the D_2^*-triangulation is the smallest one for the K_2^*-triangulation, the J_2^*-triangulation and the D_2^*-triangulation for simplicial homotopy algorithms. Furthermore, since the D_2^*-triangulation subdivides each level $\{2^{-k}\} \times R^n$, for $k = 0, 1, \cdots$, into simplices according to the D_1-triangulation of R^n, the average directional density of the D_2^*-triangulation is on each level less than the ones of the K_2^*-triangulation and of the J_2^*-triangulation. The latter two triangulations subdivide each level $\{2^{-k}\} \times R^n$ into simplices, for $k = 0, 1, \cdots$, according to the K_1-triangulation of R^n and the J_1-triangulation of R^n, respectively.

Chapter 10

Conclusions

Simplicial algorithms have extensive applications in economics, game theory, and other scientific fields, see references. Therefore, it is very significant to improve the efficiency of simplicial algorithms. As has been seen in the previous chapters, the D_1-triangulation is superior to all other triangulations of R^n underlying simplicial algorithms according to measures of efficiency of triangulations such as the number of simplices in a unit cube, the diameter, and the average directional density. Moreover, it is possible to utilize the D_1-triangulation or the variants of the D_1-triangulation in the various simplicial algorithms. By now we have implemented the D_1-triangulation to underlie the Sandwich method. Numerical tests show that based on the D_1-triangulation, the Sandwich method has become much more efficient.

The Sandwich method was discovered on the Euclidean space by Merrill in 1972 and was independently rediscovered on the unit simplex by Kuhn and MacKinnon in 1975. Let $f : R^n \to R^n$ be continuous. Our purpose is to compute an approximation of a zero point of f. In the Sandwich method a homotopy mapping h on $[0,1] \times R^n$ is defined by

$$h(t,x) = tf(x) + (1-t)(x - x^0),$$

where x^0 is an arbitrary point in R^n. Let G be a simplicial subdivision of $[0,1] \times R^n$ such that all vertices of every simplex of G are contained in

$$\{0\} \times R^n \cup \{1\} \times R^n.$$

The K_1-, J_1-, and D_1-triangulations can straightforward be utilized in the Sandwich method. Then for $\bar{x} = (t, x) \in \{0\} \times R^n \cup \{1\} \times R^n$, we define $\ell(\bar{x})$ by

$$\ell(\bar{x}) = \begin{cases} x - x^0 & \text{if } t = 0, \\ f(x) & \text{if } t = 1. \end{cases}$$

Let $\bar{G}$ denote the collection of all facets of simplices of G. An n-simplex of $\bar{G}$ with vertices $\bar{y}^0, \bar{y}^1, \cdots, \bar{y}^n$ is called complete under ℓ if there exists a nonnegative solution of the linear system

$$\lambda_0 \begin{pmatrix} \ell(\bar{y}^0) \\ 1 \end{pmatrix} + \lambda_1 \begin{pmatrix} \ell(\bar{y}^1) \\ 1 \end{pmatrix} + \cdots + \lambda_n \begin{pmatrix} \ell(\bar{y}^n) \\ 1 \end{pmatrix} = \begin{pmatrix} 0 \\ 1 \end{pmatrix}.$$

It is obvious that if $(0, x^0)$ is an interior point of some n-simplex in $\{0\} \times R^n$ then this n-simplex is a unique complete n-simplex in $\{0\} \times R^n$. For a given simplicial subdivision G with grid size δ, choose an initial point $(0, x^0)$ such that it is an interior point of some n-simplex σ in $\{0\} \times R^n$. Starting at the n-simplex σ, the Sandwich method generates a sequence of adjacent $(n + 1)$-simplices in G with common complete facets. Under the convergent condition given by Merrill in [133], the Sandwich method will terminate at a complete n-simplex σ^* with vertices $\bar{y}^0, \bar{y}^1, \cdots, \bar{y}^n$ in $\{1\} \times R^n$ within a finite number of iterations. Let $\lambda_0^*, \lambda_1^*, \cdots, \lambda_n^*$ denote the solution that corresponds to the linear system related to σ^*. Then $\sum_{i=0}^n \lambda_i^* = 1$ and

$$\lambda_0^* f(y^0) + \lambda_1^* f(y^1) + \cdots + \lambda_n^* f(y^n) = 0.$$

Let

$$x^* = \lambda_0^* y^0 + \lambda_1^* y^1 + \cdots + \lambda_n^* y^n.$$

Then x^* is an approximation of a zero point of f. When the precision is not good enough, one can restart the Sandwich method at $x^0 = x^*$ with a finer simplicial subdivision.

Let SMD1 denote the Sandwich method based on the D_1-triangulation, SMK1 the Sandwich method based on the K_1-triangulation, and SMJ1 the Sandwich method based on the J_1-triangulation. We have made computer programs of these methods in PASCAL to test numerical efficiency of these methods. Let FE denote the number of function evaluations and LP the number of linear programming pivotings. In all the following numerical tests the initial point is equal to x^0 with $x_i^0 = 0.5$ for $i = 1, 2, \cdots, n$ and the initial grid size is equal to $\delta = 0.5$. At each restart the new grid size is half the previous one. Restarting the algorithm terminates when the accuracy for $\max_{1 \le i \le n} |f_i(x^*)|$ of less than 10^{-5} has been reached. Numerical results are given as follows.

Problem A: The function $f : R^n \to R^n$ is given by

$$f_i(x) = x_i - (\sum_{j=1}^{n} x_j^3 + i)/2n$$

for $i = 1, 2, \cdots, n$. This problem can be found in Kojima and Yamamoto's [96].

Table 10.1. Numerical Results for Problem A

n	SMK1		SMJ1		SMD1	
	FE	LP	FE	LP	FE	LP
5	63	92	103	113	26	44
10	176	260	395	440	52	89
15	339	502	848	953	79	134
20	555	823	1460	1650	108	182
25	805	1203	2290	2590	142	235
30	1113	1665	3268	3703	177	288
35	1512	2244	4427	5022	215	345
40	1899	2835	5684	6464	254	403

Problem B: The function $f : R^n \to R^n$ is given by

$$f_i(x) = x_i - e^{\cos(i \sum_{j=1}^{n} x_j)}$$

for $i = 1, 2, \cdots, n$. This problem can also be found in Kojima and Yamamoto's [96].

Table 10.2. Numerical Results for Problem B

n	SMK1		SMJ1		SMD1	
	FE	LP	FE	LP	FE	LP
5	174	307	207	339	224	339
6	1316	2194	290	446	151	235
7	2854	5031	983	1854	1023	1503
8	5809	10987	1987	3955	868	1329
9	2161	3830	2640	5361	1713	2849
10	26269	48830	2556	5330	2550	4834
11			20156	59878	6163	10852
12					20268	65505

Problem C: The function $f : R^n \to R^n$ is given by

$$f_i(x) = x_i - 5\sin\left(i \sum_{j=1}^{n} x_j\right)$$

for $i = 1, 2, \cdots, n$.

Table 10.3. Numerical Results for Problem C

n	SMK1		SMJ1		SMD1	
	FE	LP	FE	LP	FE	LP
5	2802	4826	908	1713	864	1533
6	11488	19386	2508	5025	1833	3412
7			7670	14309	5645	9671

Problem D: The function $f : R^n \to R^n$ is given by

$$f_i(x) = x_i^3 - \left(n - i + 1 + \sum_{j=1}^{i} x_{n-j+1}^3\right)/n$$

for $i = 1, 2, \cdots, n$.

Table 10.4. Numerical Results for Problem D

n	SMK1		SMJ1		SMD1	
	FE	LP	FE	LP	FE	LP
5	46	86	46	101	31	36
10	100	280	100	335	55	65
15	185	605	185	725	80	95
20	295	1055	295	1265	105	125
25	430	1630	430	1955	130	155
30	590	2330	590	2795	155	185

Bibliography

[1] Allgower, E.L. and Georg, K.(1980). Simplicial and continuation methods for approximating fixed points and solutions to systems of equations. SIAM Review 22, 28-85.

[2] Allgower, E.L.,Glashoff,K., and Peitgen, H.-O.(1980). Numerical Solution of Nonlinear Equations. Lecture Notes in Mathematics 878, Springer-Verlag, Berlin.

[3] Arrow, K.J. and Hahn, F.H.(1971). General Competitive Analysis. Holden-Day, San Francisco.

[4] Barany, I.(1980). Borsuk's theorem through complementary pivoting. Mathematical Programming 18, 84-88.

[5] Broadie, M.N. and Eaves, B.C.(1987). A variable rate refining triangulation. Mathematical Programming 38, 161-202.

[6] Brooks, P.S.(1980). Infinite regression in the Eaves-Saigal algorithm. Mathematical Programming 19, 313-327.

[7] Charnes, A., Garcia, G.B., and Lemke, C.E.(1977). Constructive proofs of theorems relating to: $F(x) = y$, with applications. Mathematical Programming, 328-343.

[8] Cottle, R.W.(1982). Minimal triangulation of the 4-cube. Discrete Mathematics 40, 25-29.

[9] Cottle, R.W. and Dantzig, G.B.(1968). Complementary pivot theory of mathematical programming. Linear Algebra and Its Applications 1, 103-125.

[10] Cottle, R.W. and Lemke, C.E.(1976). Nonlinear Programming, SIAM-AMS Proceedings 9.

[11] Dai, Y., Laan, G. van der, Talman, A.J.J., and Yamamoto, Y.(1991). A simplicial algorithm for the nonlinear stationary point problem on an unbounded polyhedron. SIAM Journal on Optimization 1, 151-165.

[12] Damme, E. van(1987). Stability and Perfection of Nash Equilibria. Springer-Verlag, Berlin.

[13] Dang, C.(1991). The D_1-triangulation of R^n for simplicial algorithms for computing solutions of nonlinear equations. Mathematics of Operations Research 16, 148-161.

[14] Dang, C.(1992). The D_3-triangulation for simplicial deformation algorithms for computing solutions of nonlinear equations. Journal of Optimization Theory and Applications 75, 51-67.

[15] Dang, C.(1993). The D_2-triangulation for simplicial homotopy algorithms for computing solutions of nonlinear equations. Mathematical Programming 59, 307-324.

[16] Dang, C.(1993). The D_2^*-triangulation for continuous deformation algorithms to compute solutions of nonlinear equations. SIAM Journal on Optimization 3, 784-799.

[17] Dang, C. and Talman, A.J.J.(1990). A new triangulation of the unit simplex for computing economic equilibria. Methods of Operations Research 63, 45-56.

[18] Dang, C. and Talman, A.J.J.(1990). The D_1-triangulation in simplicial variable dimension algorithms for computing solutions of nonlinear equations. Discussion paper 9027, Center for Economic Research, Tilburg University, Tilburg, The Netherlands.

[19] Dang, C. and Talman, A.J.J.(1990). The D_1-triangulation in simplicial variable dimension algorithms on the unit simplex for computing fixed points. Discussion paper 9050, Center for Economic Research, Tilburg University, Tilburg, The Netherlands.

[20] Dantzig, G.B. and Eaves, B.C.(1974). Studies in Optimization 10. Amercian Mathematical Society.

[21] Dantzig, G.B., Eaves, B.C., and Gale, D.(1979). An algorithm for a piecewise linear model of trade and production with negative prices and bankruptcy. Mathematical Programming 16, 150-169.

[22] Day, R.H. and Robinson, S.M.(1972). Mathematical Topics in Economic Theory and Computation. Academic Press, New York.

[23] Debreu, G.(1959). Theory of Value. Wiley, New York.

[24] Doup, T.M.(1988). Simplicial Algorithm on the Simplotope. Lecture Notes in Economics and Mathematical Systems 318, Springer-Verlag, Berlin.

[25] Doup, T.M., Elzen, A.H. van den, and Talman, A.J.J.(1987). Simplicial algorithms for solving the nonlinear complementarity problem on the simplotope. In [173], 125-154.

[26] Doup, T.M., Laan, G. van der, and Talman, A.J.J.(1987). The $(2^{n+1} - 2)$-ray algorithm: a new simplicial algorithm to compute economic equilibria. Mathematical Programming 39, 241-252.

[27] Doup, T.M. and Talman, A.J.J.(1987). A new variable dimension algorithm to find equilibria on the product space of unit simplices. Mathematical Programming 37, 319-355.

[28] Doup, T.M. and Talman, A.J.J.(1987). The 2-ray algorithm for solving equilibrium problems on the unit simplex. Methods of Operations Research 57, 269-285.

[29] Doup, T.M. and Talman, A.J.J.(1987). A continuous deformation algorithm on the product space of unit simplices. Mathematics of Operations Research 12, 485-521.

[30] Eaves, B.C.(1970). An odd theorem. Proceedings of American Mathematical Society 26, 509-513.

[31] Eaves, B.C.(1971). On the basic theory of complementarity. Mathematical Programming 1, 68-75.

[32] Eaves, B.C.(1971). Computing Kakutani fixed points. SIAM Journal on Applied Mathematics 21, 236-244.

[33] Eaves, B.C.(1971). The linear complementarity problem. Management Science 17, 612-634.

[34] Eaves, B.C.(1972). Homotopies for the computation of fixed points. Mathematical Programming 3, 1-22.

[35] Eaves, B.C.(1974). Solving regular piecewise linear convex equations. Mathematical Programming Study 1, 96-119.

[36] Eaves, B.C.(1974). Properly labeled simplexes. In [20], 71-93.

[37] Eaves, B.C.(1976). A short course in solving equations with PL homotopies. In [10], 73-143.

[38] Eaves, B.C.(1978). Computing stationary points. Mathematical Programming Study 7, 1-14.

[39] Eaves, B.C.(1984). Permutation congruent transformations of the Freudenthal triangulation with minimum surface density. Mathematical Programming 29, 77-99.

[40] Eaves, B.C.(1984). A Course in Triangulations for Solving Equations with Deformations. Lecture Notes in Economics and Mathematical Systems 234, Springer-Verlag, Berlin.

[41] Eaves, B.C., Gould, F.J., Peitgen, H.O., and Todd, M.J.(1983). Homotopy Methods and Global Convergence. NATO Conference Series 13, Plenum Press, New York.

[42] Eaves, B.C. and Saigal, R.(1972). Homotopies for the computation of fixed points on unbounded regions. Mathematical Programming 3, 225-237.

[43] Eaves, B.C. and Scarf, H.(1976). The solution of systems of piecewise linear equations. Mathematics of Operations Research 1, 1-27.

[44] Eaves, B.C. and York, J.A.(1984). Equivalence of surface density and average directional density. Mathematics of Operations Research 9, 363-375.

[45] Elzen, A.H. van den and Talman, A.J.J.(1991). A procedure for finding Nash equilibria in bi-matrix games. ZOR Methods and Models of Operations Research 35, 27-43.

[46] Engles, C.R.(1980). Economic equilibrium under deformation of the economy. In [144], 213-410.

[47] Fisher, M.L. and Gould, F.J.(1974). A simplicial algorithm for the nonlinear complementarity problem. Mathematical Programming 6, 281-300.

[48] Fisher, M.L. and Gould, F.J., and Tolle, W.J.(1977). A new simplicial approximation algorithm with restarts: relations between convergence and labelling. In [87], 41-58.

[49] Forster, W.(1980). Numerical Solution of Highly Nonlinear Problems. North-Holland, Amsterdam.

[50] Forster, W.(1987). Computing "all" solutions of systems of polynomial equations by simplicial fixed point algorithms. In [173], 39-58.

[51] Freudenthal, H.(1942). Simplizialzerlegungen von beschrankter flachheit. Annals of Mathematics 43, 580-582.

[52] Freund, R.W.(1984). Variable dimension complexes Part I: Basic theory. Mathematics of Operations Research 9, 479-497.

[53] Freund, R.W.(1984). Variable dimension complexes Part II: A unified approach to some combinatorial lemmas in topology. Mathematics of Operations Research 9, 498-509.

[54] Freund, R.W.(1986). Combinatorial theorems on the simplotope that generalize results on the simplex and cube. Mathematics of Operations Research 11, 169-179.

[55] Freund, R.W. and Todd, M.J.(1981). A constructive proof of Tucker's combinatorial lemma. Journal of Combinatorial Theory (A) 30, 321-325.

[56] Fujisawa, T. and Kuh, E.(1972). Piecewise linear theory of nonlinear networks. SIAM Journal on Applied Mathematics 22, 307-328.

[57] Garcia, G.B.(1975). A fixed point theorem including the last theorem of Poincaré. Mathematical Programming 8, 227-239.

[58] Garcia, G.B.(1976). A hybrid algorithm for the computation of fixed points. Management Science 22, 606-613.

[59] Garcia, G.B.(1977). Computation of solutions to nonlinear equations under homotopy invariance. Mathematics of Operations Research 2, 25-29.

[60] Garcia, G.B. and Gould, F.J.(1978). A theorem on homotopy paths. Mathematics of Operations Research 3, 282-289.

[61] Garcia, G.B. and Gould, F.J.(1979). Scalar labellings for homotopy paths. Mathematical Programming 17, 184-197.

[62] Garcia, G.B. and Gould, F.J.(1980). Relations between several path following algorithms and local and global Newton methods. SIAM Review 22, 263-274.

[63] Garcia, G.B. and Li, T.Y.(1980). On the number of solutions to polynomial systems of equations. SIAM Journal on Numerical Analysis 17, 540-546.

[64] Garcia, G.B. and Zangwill, W.I.(1979). Determining all solutions to certain systems of nonlinear equations. Mathematics of Operations Research 4, 1-14.

[65] Garcia, G.B. and Zangwill, W.I.(1979). Finding all solutions to polynomial systems and other systems of equations. Mathematical Programming 16, 159-176.

[66] Garcia, G.B. and Zangwill, W.I.(1979). An approach to homotopy and degree theory. Mathematics of Operations Research 4, 390-405.

[67] Garcia, G.B. and Zangwill, W.I.(1980). A flex simplicial algorithm. In [49], 71-92.

[68] Garcia, G.B. and Zangwill, W.I.(1981). Pathways to Solutions, Fixed Points, and Equilibria. Series in Computational Mathematics, Prentice-Hall, Eanglewood Cliffs, NJ.

[69] Georg, K.(1979). An application of simplicial algorithms to variational inequalities. In [138], 117-127.

[70] Georg, K.(1981). A numerically stable update for simplicial algorithms. In [2], 117-127.

[71] Gould, F.J. and Tolle, J.W.(1974). A unified approach to complementarity in optimization. Discrete Mathematics 7, 225-271.

[72] Gould, F.J. and Tolle, J.W.(1976). An existence theorem for solutions to $f(x) = 0$. Mathematical Programming 11, 252-262.

[73] Gould, F.J. and Tolle, J.W.(1983). Complementary Pivoting on a Pseudomanifold Structure with Applications in the Decision Sciences. Sigma Series in Applied Mathematics 2, Heldermann Verlag, Berlin.

[74] Hansen, T.(1968). On the Approximation of a Competitive Equilibrium. PhD Thesis, Department of Economics, Yale University, New Haven, CT.

[75] Hansen, T.(1974). On the approximation of Nash equilibrium points in an N-person noncooperative game. SIAM Journal on Applied Mathematics 26, 622-637.

[76] Hirsch, M.W.(1963). A proof of the nonretractability of a cell onto its boundary. Proceedings of the American Mathematical Society 14, 364-365.

[77] Hirsch, M.W.(1979). On algorithms for solving $f(x) = 0$. Communication on Pure and Applied Mathematics 32, 281-312.

[78] Hirsch, M.W. and Smale, S.(1974). Differential Equations, Dynamic Systems, and Linear Algebra. Academic Press, New York.

[79] Hofkes, M.(1990). A simplicial algorithm to solve the nonlinear complementarity problem on $S^n \times R_+^m$. Journal of Optimization Theory and Applications 67, 551-565.

[80] Hu, T.C. and Robinson, S.M.(1980). Mathematical Programming. Academic Press, New York.

[81] Jeppson, M.M.(1972). A search for the fixed points of a continuous mapping. In [22], 122-129.

[82] John, K.(1984). Parametric fixed point algorithms with applications to economic policy analysis. Computers and Operations Research 11, 157-178.

[83] Kakutani, S.(1941). A generalization of Brouwer's fixed point theorem. Duke Mathematical Journal 8, 457-459.

[84] Kamiya, K. and Talman, A.J.J.(1991). Simplicial algorithm to find zero points of a function with special structure on a simplotope. Mathematics of Operations Research 16, 609-626.

[85] Kamiya, K. and Talman, A.J.J.(1990). Variable dimension simplicial algorithm for balanced games. Discussion paper 9025, CentER, Tilburg University, Tilburg, The Netherlands.

[86] Karamardian, S.(1972). The complementarity problem. Mathematical Programming 2, 107-129.

[87] Karamardian, S.(1977). Fixed Points: Algorithms and Applications. Academic Press, New York.

[88] Köberl, D.(1980). The solution of nonlinear equations by the computation of fixed points with a modification of the sandwich method. Computing 25, 175-179.

[89] Kojima, M.(1978). On the homotopic approach to systems of equations with separable mappings. Mathematical Programming Study 7, 170-184.

[90] Kojima, M.(1978). A modification of Todd's triangulation J_3. Mathematical Programming 15, 223-227.

[91] Kojima, M.(1978). Studies on piecewise-linear approximations of piecewise-C^1-mappings in fixed points and complementarity theory. Mathematics of Operations Research 3, 17-36.

[92] Kojima, M.(1980). An introduction to variable dimension algorithms for solving systems of equations. In [2], 199-237.

[93] Kojima, M., Nishino, H., and Arima, N.(1979). A PL homotopy for finding all the roots of a polynomial. Mathematical Programming 16, 37-62.

[94] Kojima, M. and Saigal, R.(1981). On the number of solutions to a class of complementarity problems. Mathematical Programming 21, 190-203.

[95] Kojima, M. and Yamamoto, Y.(1982). Variable dimension algorithms: basic theory, interpretation, and extensions of some existing methods. Mathematical Programming 24, 177-215.

[96] Kojima, M. and Yamamoto, Y.(1984). A unified approach to the implementation of several restart fixed point algorithms and a new variable dimension algorithm. Mathematical Programming 28, 288-328.

[97] Kremers, J.W.A.M. and Talman, A.J.J.(1990). Solving the non-linear complementarity problem. Methods of Operations Research 62, 91-103.

[98] Kuhn, H.W.(1968). Simplicial approximation of fixed points. Proceedings of National Academy of Science 61, 1238-1242.

[99] Kuhn, H.W.(1969). Approximate search for fixed points. Computing Methods in Optimization Problems 2, 199-211.

[100] Kuhn, H.W.(1974). A new proof of the fundamental theorem of algebra. Mathematical Programming Study 1, 148-158.

[101] Kuhn, H.W.(1977). Finding roots of polynomials by pivoting. In [87], 11-39.

[102] Kuhn, H.W. and MacKinnon, J.G.(1975). The sandwich method for finding fixed points. Journal of Optimization Theory and Applications 17, 189-204.

[103] Kuhn, H.W., Wang, Z., and Xu, S.(1984). On the cost of computing roots of polynomials. Mathematical Programming 28, 156-163.

[104] Laan, G. van der(1980). Simplicial fixed point algorithms. Mathematical Centre Tracts 129, Mathematisch Centrum, Amsterdam.

[105] Laan, G. van der(1982). Simplicial approximation of unemployment equilibria. Journal of Mathematical Economics 9, 83-97.

[106] Laan, G. van der(1985). The computation of general equilibrium in economies with a block diagonal pattern. Econometrica 53, 659-665.

[107] Laan, G. van der and Seelen, L.P.(1984). Efficiency and implementation of simplicial zero point algorithms. Mathematical Programming 30, 196-217.

[108] Laan, G. van der and Talman, A.J.J.(1979). A restart algorithm for computing fixed points without extra dimension. Mathematical Programming 17, 74-84.

[109] Laan, G. van der and Talman, A.J.J.(1979). A restart algorithm without an artificial level for computing fixed points on unbounded regions. In [138], 247-256.

[110] Laan, G. van der and Talman, A.J.J.(1980). An improvement of fixed point algorithms by using a good triangulation. Mathematical Programming 18, 274-285.

[111] Laan, G. van der and Talman, A.J.J.(1980). A new subdivision for computing fixed points with a homotopy algorithm. Mathematical Programming 19, 78-91.

[112] Laan, G. van der and Talman, A.J.J.(1980). Variable dimension restart algorithms for aproximating fixed points. In [49], 3-36.

[113] Laan, G. van der and Talman, A.J.J.(1981). A class of simplicial restart fixed point algorithms without an extra dimension. Mathematical Programming 20, 33-48.

[114] Laan, G. van der and Talman, A.J.J.(1981). Labelling rules and orientation: Sperner's lemma and Brouwer's degree. In [2], 238-257.

[115] Laan, G. van der and Talman, A.J.J.(1982). On the computation of fixed points in the product space of unit simplices and an application to noncooperative N-person games. Mathematics of Operations Research 7, 1-13.

[116] Laan, G. van der and Talman, A.J.J.(1983). Interpretation of the variable dimension fixed point algorithm without artificial level. Mathematics of Operations Research 8, 86-99.

[117] Laan, G. van der and Talman, A.J.J.(1983). Note on the path following approach of equilibrium programming. Mathematical Programming 25, 363-367.

[118] Laan, G. van der and Talman, A.J.J.(1986). Simplicial algorithms for finding stationary points, a unifying description. Journal of Optimization Theory and Applications 50, 262-281.

[119] Laan, G. van der and Talman, A.J.J.(1987). Ajustment processes for finding economic equilibria. In [173], 85-124.

[120] Laan, G. van der and Talman, A.J.J.(1987). Simplicial approximation of solutions to the nonlinear complementarity problem with lower and upper bounds. Mathematical Programming 38, 1-15.

[121] Laan, G. van der and Talman, A.J.J.(1987). Ajustment processes for finding economic equilibrium problems on the unit simplex. Economics Letter 23, 119-123.

[122] Laan, G. van der, Talman, A.J.J., and Van der Heyden, L.(1987). Simplicial variable dimension algorithms for solving the nonlinear complementarity problem on a product of unit simplices using a general labelling. Mathematics of Operations Research 12, 377-397.

[123] Lee, C.W.(1984). Triangulating the d-cube. IBM Thomas J. Watson Research Center Technical Report. Yorktown Heights, New York.

[124] Lemke, C.E.(1965). Bimatrix equilibrium points and mathematical programming. Management Science 11, 681-689.

[125] Lemke, C.E. and Howson, J.T.(1964). Equilibrium points of bimatrix games. SIAM Review 12, 413-423.

[126] Luthi, H.J.(1975). A simplicial approximation of a solution for the nonlinear complementarity problem. Mathematical Programming 9, 278-293.

[127] MacKinnon, J.G.(1980). Solving urban general equilibrium problems by fixed point methods. In [144], 197-212.

[128] Mangasarian, O.L.(1976). Equivalence of the complementarity problem to a system of nonlinear equations. SIAM Journal on Applied Mathematics 31, 89-92.

[129] Mansur, A. and Whalley, J.(1982). A decomposition algorithm for general equilibrium computation with application to international trade models. Econometrica 50, 1547-1557.

[130] Mara, P.S.(1976). Triangulations for the cube. Journal of Combinatorial Theory (A) 20, 170-177.

[131] Mas-Colell, A.(1985). The Theory of General Economic Equilibrium. Econometric Society Publication 9, Cambridge University Press, Cambridge, MA.

[132] Megiddo, N.(1978). On the parametric nonlinear complementarity problem. Mathematical Programming Study 7, 142-150.

[133] Merrill, O.H.(1972). Applications and Extensions of an Algorithm that Computes Fixed Points of Certain Upper Semi-Continuous Point to Set Mappings. PhD Thesis, Department of Industrial and Operations Engineering, University of Michigan, Ann Arbor, MI.

[134] Meyerson, M. amd Wright, A.H.(1979). A new and constructive proof of the Borsuk-Ulam theorem. Proceedings of American Mathematical Society 73, 134-136.

[135] Murty, K.G.(1988). Linear Complementarity, Linear and Nonlinear Programming. Sigma Series in Applied Mathematics 3, Heldermann Verlag, Berlin.

[136] Netravali, A.N. and Saigal, R.(1976). Optimal quantizer design using a fixed point algorithm. The Bell System Technical Journal 55, 1423-1435.

[137] Ortega, J.M. amd Rheinboldt, W.C.(1970). Iterative Solutions of Nonlinear Equations of Several Variables. Academic Press, New York.

[138] Peitgen, H.O.(1979). Functional Differential Equations and Approximation of Fixed Points. Lecture Notes in Mathematics 730, Springer-Verlag, Berlin.

[139] Prüfer, M. and Siegberg, H.W.(1981). Complementarity pivoting and the Hopf degree theorem. Journal of Mathematical Analysis and Applications 84, 133-149.

[140] Reiser, P.M.(1981). A modified integer labelling for complementarity algorithms. Mathematics of Operations Research 6, 129-139.

[141] Renegar, J.(1985). On the complexity of a piecewise linear algorithm for approximating roots of complex polynomials. Mathematical Programming 32, 301-318.

[142] Renegar, J.(1985). On the cost of approximating all roots of a complex polynomial. Mathematical Programming 32, 319-336.

[143] Renegar, J.(1988). Rudiments of an average case complexity theory for piecewsie-linear path following algorithms. Mathematical Programming 40, 113-163.

[144] Robinson, S.(1980). Analysis and Computation of Fixed Points. Academic Press, New York.

[145] Rockafellar, R.T.(1970). Convex Analysis. Princeton University Press, Princeton, NJ.

[146] Ruys, P.H.M. and Laan, G. van der(1987). Computation of an industrial equilibrium. In [173], 205-230.

[147] Saari, D.G. and Saigal, R.(1980). Some generic properties of paths generated by fixed point algorithms. In [144], 57-72.

[148] Saari, D.G. and Simon, C.P.(1978). Effective price mechanisms. Econometrica 46, 1097-1125.

[149] Saigal, R.(1976). On paths generated by fixed point algorithms. Mathematics of Operations Research 2, 359-380.

[150] Saigal, R.(1977). Investigations into the efficiency of fixed point algorithms. In [87], 203-223.

[151] Saigal, R.(1977). On the convergence rate of algorithms for solving equations that are based on methods of complementarity pivoting. Mathematics of Operations Research 4, 108-124.

[152] Saigal, R.(1979). The fixed point approach to nonlinear programming. Mathematical Programming Study 10, 142-157.

[153] Saigal, R.(1979). On piecewise linear approximations to smooth mappings. Mathematics of Operations Research 4, 153-161.

[154] Saigal, R.(1983). An efficient procedure for traversing large pieces in fixed point algorithms. In [41], 239-248.

[155] Saigal, R.(1983). A homotopy for solving large, sparse and structured fixed point problems. Mathematics of Operations Research 8, 557-578.

[156] Saigal, R.(1984). Computational complexity of a piecewise linear homotopy algorithm. Mathematical Programming 28, 164-173.

[157] Saigal, R., Solow, D., and Wolsey, L.A.(1975). A comparative study of two algorithms to compute fixed points over unbounded regions. In Proceedings of VII-th Mathematical Programming Symposium, Stanford, CA.

[158] Saigal, R. and Todd, M.J.(1978). Efficient acceleration techniques for fixed point algorithms. SIAM Journal on Numerical Analysis 15, 997-1007.

[159] Sallee, J.F.(1982). A triangulation of the n-cube. Discrete Mathematics 40, 81-86.

[160] Sallee, J.F.(1984). Middle cut triangulations of the n-cube. SIAM Journal on Algebra and Discrete Mathematics 5, 407-418.

[161] Samuel, A.A. and Todd, M.J.(1983). An efficient simplicial algorithm for computing a zero of a convex union of smooth functions. Mathematical Programming 25, 83-108.

[162] Saupe, D.(1982). On accelerating PL continuation algorithms by predictor-corrector methods. Mathematical Programming 23, 87-110.

[163] Scarf, H.(1967). The approximation of fixed points of a continuous mapping. SIAM Journal on Applied Mathematics 15, 1328-1343.

[164] Scarf, H.(1967). The core of an N person game. Econometrica 35, 50-69.

[165] Scarf, H.(1973). The Computation of Economic Equilibria. Yale University Press, New Haven, CT.

[166] Shamir, S.(1980). Two triangulations for homotopy fixed point algorithms with an arbitrary refinement factor. In [144], 25-56.

[167] Shapley, L.S.(1973). On balanced games without side payments. In [80], 261-290.

[168] Shoven, J.B.(1977). Applying fixed point algorithms to the analysis of tax polices. In [87], 403-434.

[169] Solow, D.(1981). Comparative computer results of a new complementarity pivot algorithm for solving equality and inequality constrained optimization problems. Mathematical Programming 18, 213-224.

[170] Solow, D.(1981). Homeomorphisms of triangulations with applications to computing fixed points. Mathematical Programming 20, 213-224.

[171] Talman, A.J.J.(1980). Variable Dimension Fixed Point Algorithms and Triangulations. Mathematical Centre Tracts 128, Mathematisch Centrum, Amsterdam.

[172] Talman, A.J.J. and Van der Heyden, L.(1983). Algorithms for the linear complementarity problem which allow an arbitrary starting point. In [41], 267-286.

[173] Talman, A.J.J. and Laan, G. van der(1987). The Computation and Modelling of Economic Equilibria. Contributions to Economic Analysis 167, North-Holland, Amsterdam.

[174] Talman, A.J.J. and Yamamoto, Y.(1989). A simplicial algorithm for stationary point problems on polytopes. Mathematics of Operations Research 14, 383-399.

[175] Todd, M.J.(1974). A generalized complementarity pivoting algorithm. Mathematical Programming 6, 243-263.

[176] Todd, M.J.(1976). The Computation of Fixed Points and Applications. Lecture Notes in Economics and Mathematical Systems 124, Springer-Verlag, Berlin.

[177] Todd, M.J.(1976). On triangulations for computing fixed points. Mathematical Programming 10, 322-346.

[178] Todd, M.J.(1976). Orientation in complementary pivot algorithms. Mathematics of Operations Research 1, 54-66.

[179] Todd, M.J.(1977). Union jack triangulations. In [87], 315-336.

[180] Todd, M.J.(1978). Improving the convergence of fixed point algorithms. Mathematical Programming Study 7, 151-179.

[181] Todd, M.J.(1978). On the Jacobian of a function at a zero computed by a fixed point algorithm. Mathematics of Operations Research 3, 126-132.

[182] Todd, M.J.(1980). A quadratically-convergent fixed point algorithm for economic equilibria and linearly constrained optimization. Mathematical Programming 18, 111-126.

[183] Todd, M.J.(1980). Exploiting structure in piecewise linear homotopy algorithms for solving equations. Mathematical Programming 18, 233-247.

[184] Todd, M.J.(1982). On the computational complexity of piecewise linear homotopy algorithms. Mathematical Programming 24, 216-224.

[185] Todd, M.J.(1984). J': A new triangulation of R^n. SIAM Journal on Algebric and Discrete Methods 5, 244-254.

[186] Todd, M.J.(1985). 'Fat' triangulations, or solving certain nonconvex matrix optimization problems. Mathematical Programming 31, 123-136.

[187] Tuy, H.(1979). Pivotal methods for computing equilibrium points: unified approach and a new restart algorithm. Mathematical Programming 16, 210-227.

[188] Van der Heyden, L.(1982). A refinement procedure for computing fixed points. Mathematics of Operations Research 7, 295-313.

[189] Varian, H.R.(1984). Microeconomic Analysis. W.W. Norton & Company, New York.

[190] Veinott, A.F. and Dantzig, G.B.(1968). Integral extreme points. SIAM Review 10, 371-372.

[191] Wacker, H.(1978). Continuation Methods. Academic Press, New York.

[192] Watson, L.T.(1979). Solving the nonlinear complementarity problem by a homotopy method. SIAM Journal on Control and Optimization 17, 36-46.

[193] Watson, L.T., Bixler, J.P., and Poore, A.B.(1989). Continuous homotopies for the linear complementarity problem. SIAM Journal on Matrix Analysis and Applications 10, 259-277.

[194] Wilmuth, R.S.(1977). A computational comparison of fixed point algorithms which used complementary pivoting. In [87], 249-280.

[195] Wright, A.H.(1981). The octahedral algorithm, a new simplicial fixed point algorithm. Mathematical Programming 21, 47-69.

[196] Yamamoto, Y.(1983). A new variable dimension algorithm for the fixed point problem. Mathematical Programming 25, 329-342.

[197] Yamamoto, Y.(1987). A path following algorithm for stationary point problems. Journal of the OR Society of Japan 30, 181-198.

[198] Zangwill, W.I.(1977). An eccentric barycentric fixed point algorithm. Mathematics of Operations Research 2, 343-359.

[199] Zangwill, W.I. and Garcia, C.B.(1981). Equilibrium programming: the path following approach and dynamics. Mathematical Programming 21, 262-289.

Vol. 325: P. Ferri, E. Greenberg, The Labor Market and Business Cycle Theories. X, 183 pages. 1989.

Vol. 326: Ch. Sauer, Alternative Theories of Output, Unemployment, and Inflation in Germany: 1960–1985. XIII, 206 pages. 1989.

Vol. 327: M. Tawada, Production Structure and International Trade. V, 132 pages. 1989.

Vol. 328: W. Güth, B. Kalkofen, Unique Solutions for Strategic Games. VII, 200 pages. 1989.

Vol. 329: G. Tillmann, Equity, Incentives, and Taxation. VI, 132 pages. 1989.

Vol. 330: P.M. Kort, Optimal Dynamic Investment Policies of a Value Maximizing Firm. VII, 185 pages. 1989.

Vol. 331: A. Lewandowski, A.P. Wierzbicki (Eds.), Aspiration Based Decision Support Systems. X, 400 pages. 1989.

Vol. 332: T.R. Gulledge, Jr., L.A. Litteral (Eds.), Cost Analysis Applications of Economics and Operations Research. Proceedings. VII, 422 pages. 1989.

Vol. 333: N. Dellaert, Production to Order. VII, 158 pages. 1989.

Vol. 334: H.-W. Lorenz, Nonlinear Dynamical Economics and Chaotic Motion. XI, 248 pages. 1989.

Vol. 335: A.G. Lockett, G. Islei (Eds.), Improving Decision Making in Organisations. Proceedings. IX, 606 pages. 1989.

Vol. 336: T. Puu, Nonlinear Economic Dynamics. VII, 119 pages. 1989.

Vol. 337: A. Lewandowski, I. Stanchev (Eds.), Methodology and Software for Interactive Decision Support. VIII, 309 pages. 1989.

Vol. 338: J.K. Ho, R.P. Sundarraj, DECOMP: an Implementation of Dantzig-Wolfe Decomposition for Linear Programming. VI, 206 pages.

Vol. 339: J. Terceiro Lomba, Estimation of Dynamic Econometric Models with Errors in Variables. VIII, 116 pages. 1990.

Vol. 340: T. Vasko, R. Ayres, L. Fontvieille (Eds.), Life Cycles and Long Waves. XIV, 293 pages. 1990.

Vol. 341: G.R. Uhlich, Descriptive Theories of Bargaining. IX, 165 pages. 1990.

Vol. 342: K. Okuguchi, F. Szidarovszky, The Theory of Oligopoly with Multi-Product Firms. V, 167 pages. 1990.

Vol. 343: C. Chiarella, The Elements of a Nonlinear Theory of Economic Dynamics. IX, 149 pages. 1990.

Vol. 344: K. Neumann, Stochastic Project Networks. XI, 237 pages. 1990.

Vol. 345: A. Cambini, E. Castagnoli, L. Martein, P Mazzoleni, S. Schaible (Eds.), Generalized Convexity and Fractional Programming with Economic Applications. Proceedings, 1988. VII, 361 pages. 1990.

Vol. 346: R. von Randow (Ed.), Integer Programming and Related Areas. A Classified Bibliography 1984–1987. XIII, 514 pages. 1990.

Vol. 347: D. Ríos Insua, Sensitivity Analysis in Multiobjective Decision Making. XI, 193 pages. 1990.

Vol. 348: H. Störmer, Binary Functions and their Applications. VIII, 151 pages. 1990.

Vol. 349: G.A. Pfann, Dynamic Modelling of Stochastic Demand for Manufacturing Employment. VI, 158 pages. 1990.

Vol. 350: W.-B. Zhang, Economic Dynamics. X, 232 pages. 1990.

Vol. 351: A. Lewandowski, V. Volkovich (Eds.), Multiobjective Problems of Mathematical Programming. Proceedings, 1988. VII, 315 pages. 1991.

Vol. 352: O. van Hilten, Optimal Firm Behaviour in the Context of Technological Progress and a Business Cycle. XII, 229 pages. 1991.

Vol. 353: G. Ricci (Ed.), Decision Processes in Economics. Proceedings, 1989. III, 209 pages 1991.

Vol. 354: M. Ivaldi, A Structural Analysis of Expectation Formation. XII, 230 pages. 1991.

Vol. 355: M. Salomon. Deterministic Lotsizing Models for Production Planning. VII, 158 pages. 1991.

Vol. 356: P. Korhonen, A. Lewandowski, J . Wallenius (Eds.), Multiple Criteria Decision Support. Proceedings, 1989. XII, 393 pages. 1991.

Vol. 357: P. Zörnig, Degeneracy Graphs and Simplex Cycling. XV, 194 pages. 1991.

Vol. 358: P. Knottnerus, Linear Models with Correlated Disturbances. VIII, 196 pages. 1991.

Vol. 359: E. de Jong, Exchange Rate Determination and Optimal Economic Policy Under Various Exchange Rate Regimes. VII, 270 pages. 1991.

Vol. 360: P. Stalder, Regime Translations, Spillovers and Buffer Stocks. VI, 193 pages . 1991.

Vol. 361: C. F. Daganzo, Logistics Systems Analysis. X, 321 pages. 1991.

Vol. 362: F. Gehrels, Essays In Macroeconomics of an Open Economy. VII, 183 pages. 1991.

Vol. 363: C. Puppe, Distorted Probabilities and Choice under Risk. VIII, 100 pages . 1991

Vol. 364: B. Horvath, Are Policy Variables Exogenous? XII, 162 pages. 1991.

Vol. 365: G. A. Heuer, U. Leopold-Wildburger. Balanced Silverman Games on General Discrete Sets. V, 140 pages. 1991.

Vol. 366: J. Gruber (Ed.), Econometric Decision Models. Proceedings, 1989. VIII, 636 pages. 1991.

Vol. 367: M. Grauer, D. B. Pressmar (Eds.), Parallel Computing and Mathematical Optimization. Proceedings. V, 208 pages. 1991.

Vol. 368: M. Fedrizzi, J. Kacprzyk, M. Roubens (Eds.), Interactive Fuzzy Optimization. VII, 216 pages. 1991.

Vol. 369: R. Koblo, The Visible Hand. VIII, 131 pages.1991.

Vol. 370: M. J. Beckmann, M. N. Gopalan, R. Subramanian (Eds.), Stochastic Processes and their Applications. Proceedings, 1990. XLI, 292 pages. 1991.

Vol. 371: A. Schmutzler, Flexibility and Adjustment to Information in Sequential Decision Problems. VIII, 198 pages. 1991.

Vol. 372: J. Esteban, The Social Viability of Money. X, 202 pages. 1991.

Vol. 373: A. Billot, Economic Theory of Fuzzy Equilibria. XIII, 164 pages. 1992.

Vol. 374: G. Pflug, U. Dieter (Eds.), Simulation and Optimization. Proceedings, 1990. X, 162 pages. 1992.

Vol. 375: S.-J. Chen, Ch.-L. Hwang, Fuzzy Multiple Attribute Decision Making. XII, 536 pages. 1992.

Vol. 376: K.-H. Jöckel, G. Rothe, W. Sendler (Eds.), Bootstrapping and Related Techniques. Proceedings, 1990. VIII, 247 pages. 1992.

Vol. 377: A. Villar, Operator Theorems with Applications to Distributive Problems and Equilibrium Models. XVI, 160 pages. 1992.

Vol. 378: W. Krabs, J. Zowe (Eds.), Modern Methods of Optimization. Proceedings, 1990. VIII, 348 pages. 1992.

Vol. 379: K. Marti (Ed.), Stochastic Optimization. Proceedings, 1990. VII, 182 pages. 1992.

Vol. 380: J. Odelstad, Invariance and Structural Dependence. XII, 245 pages. 1992.

Vol. 381: C. Giannini, Topics in Structural VAR Econometrics. XI, 131 pages. 1992.

Vol. 382: W. Oettli, D. Pallaschke (Eds.), Advances in Optimization. Proceedings, 1991. X, 527 pages. 1992.

Vol. 383: J. Vartiainen, Capital Accumulation in a Corporatist Economy. VII, 177 pages. 1992.

Vol. 384: A. Martina, Lectures on the Economic Theory of Taxation. XII, 313 pages. 1992.

Vol. 385: J. Gardeazabal, M. Regúlez, The Monetary Model of Exchange Rates and Cointegration. X, 194 pages. 1992.

Vol. 386: M. Desrochers, J.-M. Rousseau (Eds.), Computer-Aided Transit Scheduling. Proceedings, 1990. XIII, 432 pages. 1992.

Vol. 387: W. Gaertner, M. Klemisch-Ahlert, Social Choice and Bargaining Perspectives on Distributive Justice. VIII, 131 pages. 1992.

Vol. 388: D. Bartmann, M. J. Beckmann, Inventory Control. XV, 252 pages. 1992.

Vol. 389: B. Dutta, D. Mookherjee, T. Parthasarathy, T. Raghavan, D. Ray, S. Tijs (Eds.), Game Theory and Economic Applications. Proceedings, 1990. IX, 454 pages. 1992.

Vol. 390: G. Sorger, Minimum Impatience Theorem for Recursive Economic Models. X, 162 pages. 1992.

Vol. 391: C. Keser, Experimental Duopoly Markets with Demand Inertia. X, 150 pages. 1992.

Vol. 392: K. Frauendorfer, Stochastic Two-Stage Programming. VIII, 228 pages. 1992.

Vol. 393: B. Lucke, Price Stabilization on World Agricultural Markets. XI, 274 pages. 1992.

Vol. 394: Y.-J. Lai, C.-L. Hwang, Fuzzy Mathematical Programming. XIII, 301 pages. 1992.

Vol. 395: G. Haag, U. Mueller, K. G. Troitzsch (Eds.), Economic Evolution and Demographic Change. XVI, 409 pages. 1992.

Vol. 396: R. V. V. Vidal (Ed.), Applied Simulated Annealing. VIII, 358 pages. 1992.

Vol. 397: J. Wessels, A. P. Wierzbicki (Eds.), User-Oriented Methodology and Techniques of Decision Analysis and Support. Proceedings, 1991. XII, 295 pages. 1993.

Vol. 398: J.-P. Urbain, Exogeneity in Error Correction Models. XI, 189 pages. 1993.

Vol. 399: F. Gori, L. Geronazzo, M. Galeotti (Eds.), Nonlinear Dynamics in Economics and Social Sciences. Proceedings, 1991. VIII, 367 pages. 1993.

Vol. 400: H. Tanizaki, Nonlinear Filters. XII, 203 pages. 1993.

Vol. 401: K. Mosler, M. Scarsini, Stochastic Orders and Applications. V, 379 pages. 1993.

Vol. 402: A. van den Elzen, Adjustment Processes for Exchange Economies and Noncooperative Games. VII, 146 pages. 1993.

Vol. 403: G. Brennscheidt, Predictive Behavior. VI, 227 pages. 1993.

Vol. 404: Y.-J. Lai, Ch.-L. Hwang, Fuzzy Multiple Objective Decision Making. XIV, 475 pages. 1994.

Vol. 405: S. Komlósi, T. Rapcsák, S. Schaible (Eds.), Generalized Convexity. Proceedings, 1992. VIII, 404 pages. 1994.

Vol. 406: N. M. Hung, N. V. Quyen, Dynamic Timing Decisions Under Uncertainty. X, 194 pages. 1994.

Vol. 407: M. Ooms, Empirical Vector Autoregressive Modeling. XIII, 380 pages. 1994.

Vol. 408: K. Haase, Lotsizing and Scheduling for Production Planning. VIII, 118 pages. 1994.

Vol. 409: A. Sprecher, Resource-Constrained Project Scheduling. XII, 142 pages. 1994.

Vol. 410: R. Winkelmann, Count Data Models. XI, 213 pages. 1994.

Vol. 411: S. Dauzère-Péres, J.-B. Lasserre, An Integrated Approach in Production Planning and Scheduling. XVI, 137 pages. 1994.

Vol. 412: B. Kuon, Two-Person Bargaining Experiments with Incomplete Information. IX, 293 pages. 1994.

Vol. 413: R. Fiorito (Ed.), Inventory, Business Cycles and Monetary Transmission. VI, 287 pages. 1994.

Vol. 414: Y. Crama, A. Oerlemans, F. Spieksma, Production Planning in Automated Manufacturing. X, 210 pages. 1994.

Vol. 415: P. C. Nicola, Imperfect General Equilibrium. XI, 167 pages. 1994.

Vol. 416: H. S. J. Cesar, Control and Game Models of the Greenhouse Effect. XI, 225 pages. 1994.

Vol. 417: B. Ran, D. E. Boyce, Dynamic Urban Transportation Network Models. XV, 391 pages. 1994.

Vol. 418: P. Bogetoft, Non-Cooperative Planning Theory. XI, 309 pages. 1994.

Vol. 419: T. Maruyama, W. Takahashi (Eds.), Nonlinear and Convex Analysis in Economic Theory. VIII, 306 pages. 1995.

Vol. 420: M. Peeters, Time-To-Build. Interrelated Investment and Labour Demand Modelling. With Applications to Six OECD Countries. IX, 204 pages. 1995.

Vol. 421: C. Dang, Triangulations and Simplicial Methods. IX, 196 pages. 1995.